THE COMPLETE GUIDE TO HOMESTEADING

The Proven, Step-by-Step Blueprint to Self-Sufficiency and Saving Thousands Annually, Even on Less Than an Acre

Liam Armstrong

Your personal journey to self-sufficiency starts here.

Published by Authors Crib

www.AuthorsCrib.com

PREFACE

I remember the first time I stood upon my own small piece of dirt. The land stretched out, vast and slightly terrifying, yes, but overwhelmingly full of hope.

Have you felt that pull? That deep, persistent need to step away from the incessant noise? The desire to ground yourself and know, definitively, exactly where your sustenance originates? That pull is the foundation of this book.

This journey—the road to self-sufficiency—is built not on abstract ideas, but on action. Forget the sanitized theories. This is the dirt under the fingernails guide. We dive deep into the nitty-gritty realities: the spectacular failures, the frustrating missteps, and the things that, quite unexpectedly, work brilliantly.

We are going to build true self-sufficiency, not overnight, but piece by careful piece. We will dismantle the idea that this lifestyle is reserved only for those with acres of land or unlimited budgets. Even if you believe your growing space is too small, or your starting capital is too thin, I promise you: you can do this.

I wrote this manual for the pragmatist who needs proof, for the dreamer who requires a step-by-step map, and for the beginner who simply needs permission to try. Let us start building your new reality, today.

TABLE OF CONTENTS

CHAPTER ONE

THE SHOCKING TRUTH ABOUT WHY YOUR NEIGHBORS WILL ENVY YOUR PEACE (THE MENTAL LEAP FROM CONSUMER TO KING).

It starts with a feeling.

A slow, persistent discomfort you carry every time you walk the fluorescent aisles of the grocery store.

You see the gaps on the shelf. You see the price tags climbing higher than they were last week. You feel a vulnerability that is profound and unsettling.

You are a consumer. You are dependent.

This book is the blueprint for trading that vulnerability for sovereignty. It is the roadmap for the mental leap required to become the architect of your own food security. We are moving you from consumer to producer. From subject to king of your own patch of dirt.

The Shocking Truth is simple: Your neighbours will not envy your yield. They won't envy the size of your tomato harvest or the number of chickens you own. They will envy your peace. They will notice the stillness in you when the news is screaming about supply chain collapse or inflation spikes.

That peace comes from competence.

The modern world has sold us convenience at the expense of skill. We trade the ability to build, grow, and repair for the ability to click and buy. When you reclaim those skills, you reclaim internal quiet. That is the fundamental transaction we are making in this book. You are trading anxiety for action.

We must redefine success immediately. If you picture self-sufficiency as a sprawling, eighty-acre ranch requiring a massive loan and twenty hours of labor daily, you will never start. That image is a trap.

Our success is measured in smaller, more meaningful metrics:

1. The flavour of the first bean you grew yourself.

2. The satisfaction of not needing to buy eggs for a month.

3. The zero balance on your utility bill for one crucial component, like water or heat.

4. The moment you realize you created soil where only hardpan existed before.

The initial goal is not 100% self-sufficiency. That takes decades. The initial goal is 10% less dependence. If you can provide 10% of your own food, 10% of your own heat, or 10% of your own water, you have achieved a victory that is exponentially greater than the size of the number suggests. You have proven to yourself that the system is optional.

This is the mental foundation. If you can master your mind, you can master your half-acre, your quarter-acre, or your balcony. The tools and techniques that follow are useless without this core belief: You are no longer defined by what you buy. You are defined by what you create.

The shift starts today. We start small. We think vertically. We move deliberately. And we focus on resilience—the ability to keep producing food, no matter what happens outside your fence line.

WHY WE START: REVEALED: THE ONE DEEP DOUBT THAT FORCES HIGH-ACHIEVERS OFF THE GRID (AND HOW TO SILENCE THE ANXIETY PERMANENTLY)

Why are you really holding this book?

It wasn't just the price of eggs. It wasn't just the shipping news.

It was the feeling. The gnawing doubt. The deep anxiety that whispers to the high-achiever: You are not in control.

You mastered your career. You built your wealth. You optimized your time. But stand in your kitchen tonight. Ask yourself: Who feeds me? Who controls the water? Who sets the price for my family's health?

The answer is always the system. The grid. The supply chain.

We feel the fragility of it all. We recognize our dependence. This is the truth that successful, driven people cannot ignore. The realization that all our external achievements rely on a structure we cannot fix, influence, or guarantee. We become hostages to convenience.

You want to disconnect from that dependence. You need permanence.

Self-sufficiency is not about running away. It is about running toward competence. It is about trading passive consumption for active, demonstrable creation. It is the ultimate insurance policy.

We want to look at our family and know, without reservation, that the necessities are secured. We are not waiting for someone to deliver. We are the source. We are the solution.

When you plant that first seed, you are planting peace. When you secure that first water barrel, you are securing competence.

The moment you realize you can feed yourself, your deepest doubt vanishes. The anxiety about global uncertainty dissolves into the simple focus of turning soil. This is the silent revolution. This is the true independence you are seeking. We start here. We start now.

—★-★-★—

DEFINING YOUR ACRE: THE 10X YIELD ILLUSION: HOW TO SQUEEZE A FARM ONTO A BALCONY USING 'THE VERTICAL STACKS METHOD™'

YOU BELIEVE YOUR SPACE IS TOO SMALL. THAT IS THE NUMBER ONE hurdle. Forget the traditional idea of farming. We do not farm flat. We farm up.

We are implementing The Vertical Stacks Method™. This is how we defeat the size limitation.

Imagine a single parking spot. A traditional farmer parks one car there. We build a five-story parking garage. We utilize the air above the soil. That space is free. It is wasted opportunity if left empty.

Look at your deck railing. That is growing space. Look at your fence line. That is structural support.

Start thinking in cubic feet, not square feet. A four-foot by four-foot raised bed is sixteen square feet. If you build that bed three tiers high, you suddenly have forty-eight square feet of usable surface. It only uses the footprint of sixteen square feet.

This is the 10x yield illusion. The yield is real. The space used is compressed.

How do we build these vertical stacks?

Use simple, sturdy shelves. Use repurposed wooden pallets stood on end. Use PVC pipe towers drilled with holes for planting pockets.

We rely heavily on trellising. Build sturdy A-frames. Train climbing vegetables aggressively. Cucumbers, pole beans, and certain squash varieties must grow vertically. They cannot sprawl. This saves critical ground space for root vegetables and leafy greens.

Even on a balcony, you can deploy this method. Three stacked milk crates secured to a wall become a miniature garden tower. One crate holds soil. The next holds herbs. The third holds trailing strawberries.

Every single sunny wall is now a potential garden bed.

Inside the beds, we plant intensively. We do not waste space between plants. We use the Square Foot Gardening technique. We pack plants based on their mature size. The soil surface must always be covered by production.

Do not leave empty dirt. Empty dirt is lost time.

Start stacking. Start climbing. Your tiny parcel is vast once you look up.

—★—★—★—

THE BUDGET BLUEPRINT: WARNING: DO NOT BUY THESE 5 HOMESTEADING TOOLS! (THE $97 STARTER KIT THAT OUTPERFORMS THE $5,000 SETUP)

The biggest lie circulating in the self-sufficiency movement is that it requires a massive capital investment. People see glossy magazines or social media feeds showing acres of perfect rows, expensive tractors, and complex greenhouses, and they instantly assume this path is only for the wealthy. This is simply not true.

If your primary goal is to bulletproof your family's food security and save money, spending thousands on specialized equipment is counterproductive. We must prioritize needs—the ability to grow food and shelter it—over wants—the convenience of heavy machinery.

WARNING: DO NOT BUY THESE 5 HOMESTEADING TOOLS!

BEFORE YOU PULL OUT YOUR WALLET, UNDERSTAND THAT THESE FIVE items are the most common budget pitfalls for new homesteaders. Avoid them entirely until you are profitable and certain they will save you significant labor costs (which they rarely do on a micro-homestead):

1. High-End Tillers or Tractors: Unless you are turning over ten acres of field grass, hand tools are sufficient for initial bed preparation. For heavy initial work, rent a tiller for a single weekend. Your time investment in learning to use hand tools will pay off in soil health later (no compaction).

2. Advanced Hydroponic/Aquaponic Systems: These are science experiments, not immediate food security solutions. They require constant inputs, electricity, and precise pH monitoring. Master dirt first.

3. Pre-Built Chicken Coops or Runs: These are universally overpriced and often poorly designed for predator protection. We will discuss later how to build a superior, fox-proof coop using salvaged materials for a fraction of the cost.

4. Electric Dehydrators or Freeze Dryers: These consume electricity and often cost hundreds, sometimes thousands. Start with sun drying or oven drying at the lowest temperature setting. You can always upgrade later if your surplus demands it.

5. Specialized Garden Gadgets: Forget the stainless steel dibblers, expensive weeding forks, and ergonomically designed everything. Marketing targets the new gardener by offering tools for problems that don't exist.

THE $97 STARTER KIT THAT OUTPERFORMS THE $5,000 SETUP

Your budget blueprint should focus on durability and multi-functionality. This foundational kit provides everything you need to break ground, plant, and manage basic soil health:

1. A High-Quality D-Handle Spade or Shovel: This is your primary tool. Spend $30–$50 on one that feels good in your hands and has a solid wooden handle. Treat it well.

2. A Bow Rake (Metal): Essential for leveling soil, clearing debris, and blending amendments. Cost: $15–$25.

3. Pruning Shears (Bypass Style): For harvesting, trimming, and basic fruit tree care. Buy a simple, reliable pair. Cost: $10–$20.

4. Five-Gallon Buckets (3-5 minimum): The most versatile item on the homestead. Use them for carrying soil, mixing compost tea, harvesting, and temporary storage. You can often get these free from bakeries or restaurants. Assume $5 if buying new.

5. Basic Ledger and Pen: This is often overlooked but is the most important financial tool you own. Free or $2.

Total Estimated Cost: $97 (If you source well).

This minimal setup forces you to work intimately with the land, which is exactly where mastery begins.

Meticulous Records Define Success

Starting small means that every dollar saved and every ounce harvested is critical. We must approach self-sufficiency with the discipline of a CFO. From day one, you must keep meticulous records.

You cannot know if your self-sufficiency project is succeeding unless you quantify the inputs and outputs.

Your Ledger Must Track:

Seed and Material Costs: Record every purchase—seeds, soil amendments, salvaged wood, nails, feed—down to the penny.

Time Investment: Track how many hours you spend on specific tasks (planting, weeding, building the coop). This allows you to calculate the true value of your labor and identify time-wasting tasks.

Yield Measurement: Track the weight or count of everything harvested (e.g., 3.2 lbs of tomatoes, 15 dozen eggs). This is the measurable return on investment (ROI).

Why is this essential? Because when you calculate that your backyard chicken operation, after feed costs and labor, produces eggs that cost you $1.50 per dozen to raise, compared to the $4.50 store price, you have quantifiable proof of success. Homesteading is not just about peace of mind; it is about measurable economic resilience. Start tracking today.

1.4 HOMESTEADING LAWS: THE ZONING LANDMINES: 3 QUESTIONS TO ASK YOUR CITY PLANNER BEFORE YOU BUY THE CHICKENS (AVOID $10,000 FINES AND NEIGHBORHOOD WARS).

Zoning regulations are the silent killer of homesteading dreams. They represent a hidden body of law, often dense and contradictory, designed to prevent neighborhood conflicts. Ignoring them is not merely risky; it is an open invitation to financial devastation. Fines for zoning violations often start in the hundreds and rapidly escalate into the thousands, far outweighing any initial savings. Before you buy a single animal or sink a post, you must contact your local city, county, or township planning department.

Do not accept a simple "yes" or "no" answer over the phone. Always request the specific ordinance number and documentation. You need proof if a neighbor reports you later.

Here are the three mandatory questions you must ask your local authority:

1: What is the specific ordinance regarding livestock, and what are the setbacks?

Livestock is usually the biggest issue. Assume chickens are illegal until proven otherwise. Ask specifically about the number of birds allowed per property size. Often, there is a maximum limit, even in rural areas. Roosters are almost always prohibited in residential or suburban zones due to noise.

The setback requirement is critical. This is the mandated distance between the animal enclosure and your property line, or sometimes, your neighbor's dwelling. Do not guess this measurement. A coop built 15 feet from the line when the law demands 20 feet is an immediate violation and grounds for forced removal. Clarify rules for ducks, rabbits, and bees if you plan to introduce them later.

2: Are there restrictions on water catchment or greywater recycling?

Many states, and even specific counties, have stringent rules governing water use. While collecting rainwater seems like a basic right, it is often regulated. You must ask: Is a permit required to install rain barrels or cisterns? Is there a maximum allowed storage capacity?

Furthermore, clarify the rules surrounding greywater—water from sinks, showers, and laundry. Recycled greywater is a huge asset for irrigation, but it must be filtered and dispersed legally to prevent contamination and mold. Know the law before rerouting pipes.

3: At what size do agricultural structures require a building permit?

We rely heavily on sheds, compost bins, high tunnels, and cold frames. These are often considered "temporary" or "accessory" structures. Zoning departments define temporary differently.

You must determine the threshold. A cold frame or shed under a certain square footage (often 100 to 200 square feet) might be exempt from permitting. Anything larger usually requires approved plans, inspections, and fees. Building a chicken coop or a small greenhouse that exceeds this square footage without a permit can result in the city ordering you to tear it down at your own expense. Ask for the exact square footage limit and the height restriction for non-permitted structures. This prevents costly demolition later.

Your success is tied directly to your due diligence. Get the rules in writing. Only then should you start building.

CHAPTER TWO

—★-★-★—

THE 'SILENT SURVEYOR' METHOD: DESIGNING YOUR LAND FOR MAX EFFICIENCY (WHY 1 HOUR OF PLANNING SAVES 1 YEAR OF BACKBREAKING LABOR).

YOU ARE EXCITED. I KNOW THE FEELING. THE SHEER DESIRE TO GET that shovel in the dirt, to smell the freshly turned earth, and to start building your resilient future today.

Stop.

Just for an hour.

If you ignore this foundational chapter, if you rush the planning, you will spend the next five years running in circles, wasting energy, and burning yourself out doing redundant, backbreaking work. You can buy the most expensive tools and the best soil, but a poor layout means constant inefficiency. And inefficiency kills self-sufficiency dreams faster than any pest outbreak.

We are implementing the "Silent Surveyor" method. This is the process of using nothing but observation, cheap tools, and your own two feet to design a system that works for you, not against you. The goal is to maximize flow and minimize the steps you take every day.

Think of your land—whether it's a quarter acre of urban lot or five acres of pasture—as a machine. Right now, it's disorganized. We are going to engineer it. We are going to make sure that the path from the house to the compost bin is short, and the path from the rain barrel to the garden is downhill.

Here is your first real step: Grab a notepad, a pencil, and three colored pens. This initial investment in planning will return dividends measured in months of free time and thousands of saved dollars.

First, you need the base map. If you have an official survey, great. If not, draw the perimeter of your usable property as accurately as possible. Now, mark the major immovable objects: your house, your shed, large trees, the driveway, and any existing utilities (septic, water main, power poles). This is your blank slate.

Next, you are going to put on your detective hat. The Silent Surveyor phase is about asking simple questions and observing the answers over a few days, even a full cycle of seasons if you can manage the time. We are looking for the invisible forces that shape your land: the water, the sun, and the wind. If you know these forces, you can align your efforts with them, harnessing nature's power instead of fighting it.

Once you have this map complete, you hold the power to dictate exactly where the least amount of effort delivers the maximum amount of food. We are designing a property where every element assists every other element. Your chicken manure feeds the fruit trees, the fruit tree shade protects the lettuce, and the runoff from the roof waters the herbs—all without you lifting a finger or turning on a hose. That is the definition of a resilient, self-sufficient system.

Do this planning right now. Before you buy the first bag of seeds or the first plank of wood. One hour of careful visualization on paper saves 365 days of unnecessary fatigue.

MAPPING THE SUN AND WIND: HOW TO 'TRICK' YOUR GARDEN INTO BELIEVING IT'S IN THE TROPICS (MASTERING THE HIDDEN 10-FOOT MICROCLIMATE SHIFT)

THE MOST IMMEDIATE AND CRITICAL STEP IN DESIGNING AN EFFIcient homestead is learning the language of the land itself. We cannot simply look at a backyard and assume uniform conditions. Within just ten feet, you can shift your local climate one full USDA growing zone—the difference between harvesting early peppers and having them freeze in September. This is the power of the microclimate, and it starts with mapping the sun and the wind.

The truth is, even if you've lived on your land for years, you probably only know its general climate. We need to know its secrets.

Mapping the Sun: The Silent Surveyor

Sunlight is currency. We need to know exactly how much light, and what kind of heat, hits every square foot of the property throughout the growing year. This is a task that must be done in every season, not just spring.

The simplest tool you can use is a small notebook, a pencil, and your watch. For one week each season (early spring, mid-summer, early autumn), dedicate ten minutes three times a day—morning (9 a.m.), noon (1 p.m.), and late afternoon (4 p.m.)—to walk your property and observe the shadows.

1. The Thermal Mass Advantage: Note where structures like sheds, retaining walls, or large rocks absorb and radiate heat. A solid, south-facing wall (especially if it is dark brick or stone) will heat up significantly during

the day and release that warmth slowly into the night. This area, usually within a 5-foot radius of the wall, is your thermal microclimate. It is the perfect spot for plants that need extra heat to thrive, such as basil, tomatoes, or exotic peppers.

2. The Shade Line: Track how high the sun gets in mid-summer versus late autumn. A vegetable garden that gets full sun in June might be completely shaded by a neighbor"s house or a deciduous tree by September. This realization informs where you should place your annuals (which need summer sun) versus your perennial fruit trees (which can handle late-season shade).

3. Cold Pockets (The Sun Trap Failure): Notice areas that receive direct morning sun but are blocked in the afternoon, or vice versa. These spots can lead to uneven heating or, worse, become cold sinks where frost lingers longer than elsewhere, damaging tender transplants.

Mastering the Hidden 10-Foot Microclimate Shift

The real magic of the microclimate is using this data to create artificial tropical havens. If your region has a short growing season, planting sensitive crops next to a south-facing thermal mass can effectively extend your season by two weeks on either end.

—★–★–★—

MAPPING THE WIND: IDENTIFYING THE ENERGY DRAIN

Wind is the great desiccator. It strips moisture from leaves, pulls heat away from soil, and physically damages plants. Mapping the wind allows you to place protective barriers exactly where they are needed.

1. The Dominant Flow: Hang simple, colored ribbons or strips of flagging tape on existing structures or stakes around your property. Observe them for a few days, noting the direction of the strongest, coldest winds

(usually the winter wind). This dominant direction dictates where you must build your windbreaks.

2. The Windbreak Effect: A good windbreak (a fence, a dense hedge, or a row of fast-growing shrubs) does not stop the wind; it lifts it up and over the protected area. Crucially, the shelter offered extends roughly ten times the height of the windbreak. If you build a 6-foot fence, you gain 60 feet of protection directly downwind.

3. The Cold Sink Identification: Wind often funnels cold air into specific low-lying areas, creating "frost pockets." If you notice dew or frost remaining in one spot hours after it has melted everywhere else, mark that spot. This area should be reserved for hardier plants (like root vegetables) or structures (like a shed), never for your sensitive crops.

By meticulously tracking the movement of sun and air, you stop fighting the environment and start cooperating with it. You are no longer gardening in your generic geographic location; you are gardening in the unique, highly favorable microclimate you have engineered. This is the difference between struggling for meagre yields and harvesting abundance.

ZONES OF INTENSITY: THE 5-STEP FOOT TRAFFIC OPTIMIZATION HACK: PUTTING YOUR EGGS, HERBS, AND COMPOST WHERE THEY BELONG (VALUE: $1,200 IN SAVED TIME)

THINK ABOUT HOW MANY STEPS YOU TAKE IN A YEAR SIMPLY MOVING things. Carrying kitchen scraps to the compost pile. Walking to the far side of the yard to gather eggs. Fetching basil from a patch hidden behind the garage. These micro-trips are the silent killers of momentum. They introduce friction into your routine. Over a year, they add up to days of wasted labor.

The Foot Traffic Optimization Hack is a simple organizational strategy based on one rule: Place items according to the frequency of use. If you use something ten times a day, it needs to be an arm's length away. If you use it ten times a year, it can be at the edge of your property.

This strategy saves time, reduces effort, and vastly improves your chances of sticking with the routine, because the hard work is made convenient.

—★-★-★—

THE 5-STEP OPTIMIZATION HACK FOR MAXIMUM EFFICIENCY

STEP 1: THE DAILY GRAB (THE IMMEDIATE HUB)

THIS AREA ENCOMPASSES THE SPACE IMMEDIATELY OUTSIDE YOUR most-used back door, extending roughly 10 to 20 feet. Since you walk through this space countless times a day, everything here should relate to daily consumption and quick disposal.

What belongs here? High-value herbs (chives, parsley, mint, thyme) for cooking, a small salad green bed (quick harvests), and your dedicated kitchen-scrap compost pail or worm farm. You should be able to step out, grab what you need, and return without taking more than ten steps.

STEP 2: THE FREQUENT ROUTINE (THE TWICE-A-DAY PATH)

THIS AREA IS SLIGHTLY FURTHER OUT, MAYBE 20 TO 50 FEET FROM the house. This path is travelled with purpose, typically once in the morning and once in the evening. It needs to feel convenient, but it can accommodate structures that have their own smells or noise.

What belongs here? The chicken coop. You need access for feeding and egg collection. The small greenhouse or cold frame for seed starting and protecting high-value plants. The main firewood rack if you heat with wood. The key is that the path is direct and clear.

STEP 3: THE REGULAR MAINTENANCE (THE WEEKLY COMMITMENT)

This zone extends past the routines of the house and is reserved for things that need dedicated, but not daily, attention. Think about tasks that happen weekly: weeding, turning soil, or large-batch harvesting.

What belongs here? Your main raised bed system for annual crops (tomatoes, squash, beans, potatoes). The large, active composting piles that require weekly turning or aeration. If you have small livestock that only require weekly cleaning (like rabbits), their housing fits here.

STEP 4: THE OCCASIONAL CHECK-IN (THE DEEP EDGE)

This area requires attention primarily during specific seasons (planting, pruning, harvest) and may extend toward the edges of your property line. The land here can be used for passive systems that maintain themselves for long periods.

What belongs here? Perennial berry patches (raspberries, blueberries) that only need pruning once a year. Small fruit trees that are established and only require occasional watering. Low-maintenance structures like an underground root cellar or a deep water storage tank.

STEP 5: THE WILD EDGE (THE BOUNDARY/LONG-TERM RESOURCES)

This is the area furthest from the house, usually at the property boundary, and serves as a buffer. It is used for items that require the least amount of active management.

What belongs here? Timber or large tree systems managed for future firewood or lumber. Areas reserved for foraging or natural habitat. If you plan to graze larger animals (goats, sheep) in the future, their pasture entrance begins here.

By organizing your property this way, you remove the constant friction points that lead to procrastination and burnout. You ensure that the most rewarding, high-yield tasks—like grabbing fresh herbs for dinner—are effortless, reinforcing the joy of the process. Efficiency, in a small homestead, is not a luxury; it is the difference between success and failure.

—★-★-★—

BUILDING BASICS: SHEDS, BEDS, AND TRELLISES: THE 3 ESSENTIAL STRUCTURES YOU MUST ERECT IN THE FIRST 14 DAYS (USING 90% SALVAGED WOOD)

YOU HAVE SECURED YOUR MINDSET (CHAPTER 1) AND MAPPED YOUR land (Chapter 2). Now we move from theory to action. The first two weeks on your property are critical. You must define space and secure your basic assets. This is not about building a quaint, Instagram-ready farm. This is about building function using 90% salvaged materials.

The three essential structures you must erect immediately are: the raised bed, the simple trellis, and the storage shed. These structures define your work area, contain your valuable soil, and protect your tools.

The Foundation: Simple Raised Beds

Why build raised beds immediately? First, they define your soil. If you are starting on lawn or poor ground, the raised bed allows you to immediately import good soil and compost, creating a perfect growing environment without the time needed to rehabilitate the native earth. Second, they are fast.

We are aiming for a functional box. Do not buy expensive cedar lumber. Instead, source:

Salvaged Materials Checklist:

1. Discarded decking or fence boards (check local demolition sites or classifieds).

2. Untreated pallets (look for the "HT" stamp, meaning heat-treated, not chemically treated).

3. Concrete blocks (if available cheaply).

The easiest starter bed is a 4x8 foot rectangle. Four feet is the maximum width you can reach across without stepping on the soil. Aim for a depth of 6 to 12 inches.

Construction Quick-Start: Disassemble your salvaged boards. Use screws (this is one place you should invest money, as screws hold better than nails in rough wood) to join the boards into a simple rectangle. Brace the corners using interior blocks if the wood is thin. Once constructed, set the bed directly on the ground. Fill it with a mixture of scavenged yard waste, collected leaves, and whatever bagged topsoil/compost you can afford to start. Your ongoing composting efforts will fill the rest later (Chapter 3).

THE EFFICIENCY MULTIPLIER: THE QUICK TRELLIS

Space is money. If you let vining plants like cucumbers, squash, peas, or pole beans sprawl on the ground, they take up tremendous space, are prone to rot, and attract pests. Trellising is non-negotiable for maximizing yield in small spaces.

We need vertical support, and we need it now.

Salvaged Materials Checklist:

1. Scrap pieces of metal fencing (hog panel or cattle panel, if free or very cheap).

2. Straight, sturdy branches collected from nearby woods or recent tree trimming.

3. Disassembled pallet stringers (the thicker pieces of a pallet).

The most beginner-friendly and space-efficient design is the A-frame trellis. Cut two pieces of salvaged wood or two thick branches to 6 feet long. Hinge them together at the top using salvaged rope or wire. Spread the bottom edges out so the trellis stands firmly over the center of your new raised bed. For the structure, use salvaged chicken wire, simple garden twine, or cheap netting strung horizontally between the legs. The goal is to give the plant something to grab. This structure costs almost nothing and will double the yield of that raised bed instantly.

—★–★–★—

THE PROTECTIVE SHELL: THE BASIC MICRO-SHED

THE ABSOLUTE FIRST PRIORITY IS TOOL SECURITY AND MATERIAL storage. Leaving tools exposed means rust and replacement costs. You must have a dry, secure space for your essential tools and, crucially, for the dry salvaged materials you continue to collect.

This initial shed does not need to be fancy or meet building code (check your zoning, Chapter 1.4, for size limits). We are building a basic 4x6 foot storage box with a roof.

Salvaged Materials Checklist:

1. Pallets (for the floor and walls).

2. Scrap plywood (for walls and sheathing).

3. Old corrugated tin or plastic roofing panels (essential for a waterproof roof).

Construction Quick-Start: Build a simple, pallet-based floor platform slightly raised off the ground. Erect four corner posts using the thickest scrap wood you have. For the roof, ensure a slight slope to drain water away from the door. Scour local construction sites or classifieds for discarded asphalt shingles or, ideally, pieces of metal roofing. Waterproofing is the single most important feature of the shed. Secure the walls using scrap plywood or pallet wood nailed vertically. A simple latch on a salvaged door keeps your initial tool investment safe from theft and weather. This shed is the hub of your operation for the first year. Build it quickly and make it dry.

——★-★-★——

FENCING AND BARRIERS: THE LOW-COST, DEER-PROOF 'INVISIBLE WALL' SYSTEM THAT FOILS RACCOONS AND STOPS RABBITS DEAD IN THEIR TRACKS

THE REALITY OF GROWING FOOD IS SIMPLE: IF IT IS GOOD ENOUGH for you, it is good enough for every hungry creature in your area. Pests will come. They are not malicious; they are just efficient opportunists. A successful homestead treats its garden not as a farm, but as a fortress. Proactive defence is always cheaper, in both time and money, than replacing a lost harvest.

The key to low-cost barrier success is layering your defences and understanding the specific threat you face.

THE GROUND DEFENCE: STOPPING RABBITS AND VOLES

RABBITS ARE NOTORIOUS FOR NIPPING YOUNG PLANTS AT THE BASE, and they are masters of finding tiny gaps. Voles and gophers are silent destroyers, attacking roots from below the surface.

To defeat the digging pests, the fence must not stop at ground level.

1. The Rabbit Barrier: Use galvanized wire mesh, ideally 1/2-inch to 1-inch diameter (not flimsy chicken wire, which can be chewed or stretched). The barrier must extend at least 6 inches below the soil line. For absolute security, bend the bottom 4 inches of the wire into an "L" shape facing outward (away from the garden). When a rabbit tries to dig, it hits this apron and gives up.

2. The Vole Fortress: If you are using raised beds, you must line the entire bottom interior with 1/4-inch hardware cloth before adding soil. This is a non-negotiable step. Hardware cloth is tightly woven and robust enough to prevent voles and gophers from tunnelling up into your root crops.

THE JUMPING DEFENCE: FOILING DEER

DEER ARE MASSIVE ANIMALS CAPABLE OF CLEARING ANYTHING UNDER six feet. Building a perimeter fence high enough to stop them (eight feet or more) can be prohibitively expensive and visually obtrusive. This is where the "Invisible Wall" system comes into play.

Deer rely heavily on their eyesight and depth perception. They will not jump a barrier they cannot clearly gauge, especially if they are unsure of the landing zone.

THE INVISIBLE WALL METHOD

Set up a simple perimeter using sturdy T-posts and inexpensive woven wire or even simple polypropylene mesh fencing (at least four feet high). This is your visible, secondary boundary.

Now, 3 to 4 feet outside of this main fence, string two or three parallel lines of heavy-duty, monofilament fishing line (40 to 60 lb test is fine) or thin black electrical fence wire. Place one strand at 2 feet high and the other at 4 feet high.

When the deer approaches, it is confronted by two confusing visual lines separated by a gap. The deer often mistakes the two separate lines for a single, deep barrier, and the slight, surprising friction of the fishing line hitting its chest is enough to deter it from jumping over the inner fence. It works because it messes with their perception, not just their physical ability.

THE CLIMBING DEFENSE: THE RACCOON AND OPOSSUM SOLUTION

Raccoons are perhaps the most persistent and intelligent garden raiders. They love sweet corn and ripe tomatoes and will climb almost anything.

The solution requires a small amount of low-voltage electricity. Do not be intimidated by the idea of electric fencing; the systems used for backyard pest control are low-amperage, harmless, and surprisingly affordable.

Install a single strand of electric fence wire about 4 inches off the ground, mounted on small insulators attached to your existing fence posts. Power it with a small solar charger, often sold as a "paddock charger" for horses. When the raccoon, moving low to the ground, touches the wire, it receives a mild, memorable zap. This is a behavioural training tool. After one or two attempts, the raccoon learns that your fortress is protected, and they will move to easier prey. This defence is cheap, effective, and nearly invisible.

By combining the buried apron (rabbits/voles), the perceptual outer barrier (deer), and the low-level electric wire (raccoons/opossums), you create a layered system of defence that defeats the vast majority of threats with a relatively small investment in materials.

CHAPTER THREE

THE 3 POUND TOMATO SECRET: UNLOCKING SOIL FERTILITY (WHY 90% OF GARDENERS FAIL BEFORE THEY EVEN PLANT A SEED).

We come now to the heart of the matter. We talk about self-sufficiency, about cutting the cord, about building resilience. But all of that is theoretical until you put your hands into the dirt. Until you connect with the living medium that feeds you. The soil is not just filler. It is the bank account of your homestead, the vital ecosystem that determines whether your work yields abundance or failure.

The brutal truth is this: ninety percent of gardening failures—the spindly plants, the low yields, the endless battles with pests and disease—are rooted in poor soil preparation. People rush the soil work to get to the planting, and they pay for that impatience for the entire season. You can have the best seeds, the perfect sunlight, and all the water you need, but if your soil is dead, hard, or chemically imbalanced, you are failing before the journey even begins.

We are aiming for the three-pound tomato. That massive, sun-warmed fruit that smells like summer and needs two hands to hold. That tomato is not an accident. It is the physical manifestation of perfectly balanced fertility. It represents a living connection between your effort and the earth's overwhelming capacity for production.

The shift in mindset required here is simple but profound: Stop trying to feed the plants, and start feeding the soil.

When you spend an afternoon amending a bed—digging in aged manure, turning in leaf mold, smelling that rich, dark, fungal scent—you are not doing maintenance. You are engaging in an ancient conversation. There is a deep, quiet satisfaction in knowing that you are creating life. This is where the anxiety about future fragility dissolves, replaced by the certainty of a hand full of dark, fertile earth.

This chapter is your blueprint for maximizing that fertility in the smallest space possible. We will learn how to read the silent language of your dirt, how to turn simple kitchen scraps into rocket-fuel fertilizer, and how to set up your growing areas for overwhelming, relentless productivity. We will turn your garden patch, balcony, or backyard into a place of profound abundance, making every hour of your labor count ten times over. This is the bedrock of your independence. Let's get to work unlocking the secret life beneath your feet.

SPEAKING TO THE SOIL: THE 'SNIFF TEST' AND THE 'SQUEEZE TEST': TWO BACKYARD TRICKS TO REPLACE $300 LAB REPORTS AND GET PERFECT PH

THE REAL FACTORY, THE ENGINE OF THE ENTIRE OPERATION, IS THE soil. It is a living, breathing microbiome. If you treat it like inert dirt, it will reward you with inert results. If you treat it like a thriving, complex ecosystem, the payoff is immediate and massive. You need to learn how to speak its language. You don't need a university lab to know if your soil is happy. You just need your hands and your nose.

THE SNIFF TEST: IS YOUR SOIL ALIVE?

Healthy soil has a distinctive, pleasing odor. It should smell like the deep woods after a spring rain. This rich, sweet, earthy aroma is caused by beneficial bacteria called actinomycetes. These microbes thrive in balanced, aerobic (oxygen-rich) conditions. If your soil smells sour, rancid, or faintly metallic, that indicates too much water, a lack of oxygen, or an excess of undecomposed organic matter. Sour soil is sick soil. It needs aeration and carbon matter. Healthy soil smells alive.

THE SQUEEZE TEST: WHAT IS YOUR SOIL'S DAY JOB?

This test tells you instantly about the physical structure, drainage, and water retention capacity of your land. Take a handful of damp—but not dripping wet—soil from about six inches down.

1. The Squeeze: Squeeze it hard in your palm.
2. The Release: Open your hand and observe the soil ball.

If the soil holds a tight, sticky ball and refuses to break apart, you have heavy clay. Clay compacts easily and drowns roots. If the soil forms a loose, crumbly mass that immediately disintegrates when you gently poke it, you have sandy soil. Sand drains too fast and loses nutrients quickly. The ideal result is a soft, cohesive ball that holds its shape but crumbles easily when gently tapped. This structure is called loam. It holds water just long enough and provides plenty of space for root growth.

THE LOW-TECH PH TEST

You can skip the $300 lab report for a basic pH assessment. All you need is white vinegar and baking soda. pH is critical because it dictates how well your plants can access nutrients.

1. Sample One (Testing for Alkalinity): Place a small soil sample in a container. Add a splash of distilled water and then a large spoonful of baking soda. If the soil fizzes or bubbles, it is acidic (low pH).

2. Sample Two (Testing for Acidity): Place a separate soil sample in another container. Add a few tablespoons of white vinegar. If this soil sample fizzes, it is alkaline (high pH).

If neither sample fizzes, you are likely in the neutral zone, which is perfect for most common vegetables.

NATURAL AMENDMENTS FOR REVITALIZATION

Once you know the issue, the corrections are simple and organic.

Heavy Clay: This needs structural improvement. Never till it when wet. The universal remedy is to mix in massive amounts of finished compost and aged wood chips. Compost adds structure that opens up the clay, allowing for drainage and air flow. Coarse sand can help, but avoid fine sand, which just mixes with the clay to make concrete.

Sandy Soil: This needs retention power. Add heavy, water-retentive organic matter. Use well-rotted manure, leaf mold, and compost. These materials act like sponges, holding moisture and nutrients right near the root zone.

Acidic Soil (low pH): Most vegetables prefer slightly acidic to neutral soil (6.0 to 7.0). If you are too acidic, add agricultural lime or hardwood ash slowly over time.

Alkaline Soil (high pH): If your soil is too alkaline, you can drop the pH by adding elemental sulphur. Coffee grounds and peat moss are effective, quick-acting acidifiers, but use peat sparingly.

Remember: Compost fixes everything. It is the single greatest soil amendment, regardless of the diagnosis. Focus 90% of your effort on creating and applying high-quality compost, and your soil will reward you with explosive growth.

—★–★–★—

INTENSIVE GARDENING: THE SQUARE-FOOT CONSPIRACY: HOW TO HARVEST 100 HEADS OF LETTUCE IN THE SPACE OF A PARKING SPOT (THE URBAN FARMING ADVANTAGE)

We don't have room to waste. Wide, sprawling rows are inefficient. They belong on massive farms. Your goal is maximum yield in minimum space. This requires intensive planting methods.

The square-foot system is revolutionary. It redefines garden density. Traditional rows waste space on paths. Square-foot gardening eliminates those paths entirely.

You build raised beds. Make them 4 feet by 4 feet. This size is easy to manage. You never step on the soil. This prevents compaction.

Next, divide the bed into 16 squares. Use string or small wooden slats. Each one-foot square is a micro-plot. Planting density changes for each crop.

Small items plant densely. Radishes can go 16 per square. Carrots fit 9 to 16 per square. Medium items fit less. Lettuce is four plants per square. Larger items require the entire square. One head of cabbage uses one square. A single tomato plant takes one square. You get immediate visual feedback. You see exactly how much you can grow.

Yield is maximized by focusing on high-value crops. Forget corn. Grow nutrient-dense greens.

We must also utilize the third dimension. Space is not just flat ground. It is vertical. Vertical gardening drastically increases effective acreage.

Build simple structures. Use A-frame trellises. They are stable and easy to construct. Use salvaged cattle fencing or wire mesh.

Plant climbing vegetables near the trellis base. Pole beans climb fast. Cucumbers sprawl upwards. Winter squash can be trained vertically. This gets the heavy fruit off the ground.

Use stacked containers for herbs. Old tires or wooden pallets work well. Paint them dark for heat absorption. Grow strawberries in vertical PVC pipes. Cut holes along the length. Plant one berry plant per hole. This provides hundreds of berries in a 3-foot footprint.

Always plant on the north side of the garden first. The tallest items go there. This prevents shading shorter crops.

When one crop finishes, replant immediately. Do not leave the square empty. Sequential planting ensures continuous harvest. Use fast-growing crops like spinach or bush beans for quick turnover. Intensive gardening demands consistency. Every single square foot must produce food constantly.

COMPOSTING MASTERY: DISCOVERED: THE '2-WEEK HOT COMPOST' HACK THAT TURNS KITCHEN SCRAPS INTO $10/BAG FERTILIZER OVERNIGHT

COMPOST IS THE SINGLE GREATEST TOOL IN THE SELF-SUFFICIENCY arsenal. It is alchemy. You take kitchen scraps, paper, and yard waste—things you would otherwise pay to have removed—and transform them into a biologically rich, valuable soil amendment. This transformation saves you money on store-bought inputs and makes your garden yield exponentially better. Composting is the clearest example of waste transformed into wealth on the small homestead. We need two core skills here: speed composting for immediate volume, and worm composting for liquid gold.

THE 2-WEEK HOT COMPOST HACK

HOT COMPOSTING IS ABOUT CONTROLLED DECOMPOSITION. WE ARE creating a furnace. We are feeding bacteria so efficiently they generate internal temperatures between 131°F and 160°F. This heat is crucial: it kills weed seeds, pathogens, and accelerates the breakdown process. This is the method that gets you usable compost in two weeks instead of six months.

The secret is the ratio. Forget complex chemistry. Think of your materials in two categories, aiming for roughly a 30:1 ratio of Carbon to Nitrogen (C:N).

Nitrogen (Greens): These are wet, dense, and protein-rich. Examples include fresh grass clippings, kitchen scraps (fruit/veggie peels), coffee grounds, and fresh manure.

Carbon (Browns): These are dry, airy, and structural. Examples include dry leaves, shredded cardboard, newspaper, sawdust, and straw.

The Simple Ratio Rule: Build your pile using one bucket of green material for every two to three buckets of brown material. Crucially, mix thoroughly as you layer.

The Critical Steps for Speed:

1. Water: The pile must feel like a wrung-out sponge. If it is too dry, the bacteria go dormant. If it is too wet, the pile goes anaerobic and starts to smell sour.

2. Size: The pile must be at least three feet by three feet by three feet (a cubic yard) to generate the necessary thermal mass to heat up.

3. The Turning Schedule: This is the hack. Once the pile is built, it will heat up rapidly. When the internal temperature peaks (usually days 2–4), turn the entire pile. Physically move the material from the outside edges into the center. Turn it every other day after the initial turn. Each turn reoxygenates the bacteria, reignites the heat, and dramatically accelerates the process. If you commit to this turning schedule, you will have dark, finished compost ready for the garden in 14 to 21 days.

VERMICOMPOSTING: BLACK GOLD IN A BOX

For small spaces, balconies, or specialized seed starting mixes, nothing beats worm castings. Vermicomposting uses specific earthworms—usually Red Wigglers (Eisenia fetida)—to digest organic waste. Their output, called castings, is a highly concentrated, biologically complete, and nutrient-dense fertilizer.

Setting Up Your Bin: You need a dark, shallow container (10–20 gallons is a great starter size) with drainage holes. Fill it with bedding—shredded newspaper, coir, or aged compost. The bedding must be kept consistently damp.

Feeding the Worms: Worms eat slowly. Start with a small handful of soft, non-acidic scraps: melon rinds, pumpkin, coffee grounds, and tea bags. Avoid meat, dairy, oil, and citrus entirely, as these will attract pests, breed harmful bacteria, and hurt the worms.

The Real Secret Weapon: Worm Tea (Leachate): As water drains through the bedding and the rich castings, it creates a powerful liquid fertilizer. Dilute this liquid until it looks like weak tea and use it as a powerful foliar spray or root drench. This concentrated nutrient liquid is arguably the most valuable and easy-to-harvest product your composting system will produce.

Composting removes you from the cycle of dependence. You close the loop on your waste stream, stopping yourself from buying expensive inputs and instead, creating them. This is essential self-sufficiency.

—★-★-★—

SEED STARTING AND SEASON EXTENSION: HOW TO CHEAT WINTER OUT OF 3 MONTHS: THE $19 COLD FRAME SECRET THAT GETS YOU FRESH GREENS IN FEBRUARY

THE CALENDAR IS A LIMITATION IMPOSED BY NATURE. BUT WE ARE homesteaders. We rewrite the rules of the season using simple physics and careful timing. The ability to harvest a head of kale in February—three months before the official start of spring—is the ultimate sign of resilience. This is achieved through starting early and ending late.

Getting a Jump Start: Indoor Seedlings

The single biggest mistake a new gardener makes is waiting for the last frost date to plant seeds. We must gain weeks, often months, by starting indoors. This allows us to transplant mature, robust seedlings into the garden when the soil is ready, immediately maximizing their growth potential.

You do not need an expensive greenhouse. You need three things: light, drainage, and consistency.

1. The Light Solution: Forget specialized, expensive grow lights. Simple LED shop lights, available for $15-$25 at any hardware store, work perfectly. Suspend them on chains or adjustable hooks so they hang 2 to 3 inches directly above your young seedlings. Keep them on a timer for 14 to 16 hours per day. The intensity, achieved by keeping the lights low, is the key to strong, stocky plants, not leggy, weak ones.

2. The Container Secret: Start small. Recycled containers—yogurt cups, toilet paper rolls cut into segments, or peat pots—are ideal. Ensure every single container has drainage holes. Never let your seedlings sit in stagnant water.

3. The Schedule: Use an almanac or online calculator to determine your average last frost date. Count backward 6 to 8 weeks for slow growers like tomatoes, peppers, and broccoli. Count back 4 weeks for fast crops like squash or cucumbers. Starting too early can lead to root-bound plants that struggle upon transplant.

THE $19 COLD FRAME SECRET

Once the calendar says winter, the fight is not over. We transition our harvest to protected environments, primarily low-cost, passive solar structures known as cold frames or mini hoop houses.

The cold frame operates entirely on solar gain and insulation. Think of it as a small, unheated greenhouse that sits directly over a raised bed or ground plot. It allows you to continue harvesting cold-hardy crops like spinach, kale, bok choy, and scallions throughout the winter in all but the deepest freezes.

Building the Frame:

The simplest cold frame can be constructed for minimal cost:

1. The Walls: Use four pieces of salvaged 2x6 lumber or bricks to create a rectangular box directly on the ground. For improved light capture, the back (north side) should be slightly taller than the front (south side).

2. The Lid: This is the critical piece. Source an old window (check salvage yards or classifieds), or staple clear 6-mil plastic sheeting to a simple wooden frame. Hinge this lid to the taller back wall.

3. The Operation: The frame should be opened (vented) on sunny days when temperatures rise above 40 degrees Fahrenheit to prevent the plants from overheating and rotting. It must be closed tightly before sunset to trap accumulated heat and protect the plants from frost.

By employing this simple, passive technology, you create a microclimate 10 to 20 degrees warmer than the outside air. This allows those cold-hardy greens to continue slow, steady growth, providing fresh winter harvests when the grocery store produce is bland and expensive.

HARDENING OFF: THE CRITICAL TRANSITION

When your indoor seedlings are ready for the outdoors, they cannot be simply tossed into the garden. They have lived a pampered, climate-controlled life. This is where 'hardening off' comes in.

This process prepares the plants for sun, wind, and temperature shifts, preventing immediate shock and death. Begin 7 to 10 days before final planting:

Day 1-3: Place seedlings outside in a shaded, protected location for 2-4 hours.

Day 4-6: Move them to a spot receiving morning sun and leave them out for 6-8 hours. Bring them in at night.

Day 7-10: Leave them out overnight, provided temperatures stay above freezing. Introduce them to light wind and full sun exposure for increasing periods.

After this critical ten-day adjustment, your seedlings will be tough, resilient, and ready to face the world—and deliver a maximum yield during your compressed growing season. You have successfully cheated winter.

CHAPTER FOUR

—★-★-★—

THE GRID-FREE SECRET: HOW TO DITCH YOUR UTILITY BILL FOREVER USING THE 'TINY ARRAY' SYSTEM (WATER AND ELECTRICITY SOLVED).

THE UTILITY COMPANY SENDS YOU A BILL EVERY MONTH. IT'S A QUIET tax on your life. A reminder that they hold the switch. They control the flow.

We dealt with the fragility of the food system in the first three chapters. Now we look at the two other systems that hold us hostage: water and electricity. Breaking free from utility dependence is not just about saving money, though the savings are massive. It is about removing vulnerability.

Start with water. It is the most vital asset on your land. A few dry weeks can wipe out your garden and stress your animals. Relying on a municipal source or an electric well pump tied to the grid is a huge risk. Water is life. We are going to capture the rain, filter it simply, and recycle the wastewater we use every day. You are literally letting thousands of gallons of clean, free water fall off your roof every year and vanish into the storm drain. That ends today. We will turn the sky into your utility.

Then there is power. People look at going off-grid and immediately think they need to spend fifty thousand dollars on a gigantic solar field and a wall of batteries. You do not. That is the grid-scale thinking applied to the homestead. We are minimalists here. We are focused on critical loads.

What do you absolutely need when the lights go out?

A way to pump water.

A way to keep preserved food frozen.

A light to see by.

A place to charge essential communications.

That is the scope of the "Tiny Array" system. It is a small, modular setup designed to handle just those critical elements. We break down the intimidating math into simple steps. We figure out exactly how much power your well pump needs, and we buy just enough panel and battery to cover that. We are aiming for resilience, not a lifestyle upgrade.

Security is knowing that a massive blackout won't impact your ability to hydrate your family, feed your livestock, or cook a meal. We are building that security now. This chapter guides you through the process of setting up these two essential systems—water and power—simply, safely, and affordably. You are not building a nuclear reactor. You are building freedom.

HARVESTING THE SKY: THE $49 DOWNSPOUT BYPASS: COLLECTING 300 GALLONS OF PURE, FREE WATER IN A SINGLE STORM (FILTER SAFETY CRITICAL)

LET'S LOOK UP. THAT MASSIVE, ANGLED SURFACE ABOVE YOUR HEAD—your roof—is a water collection machine. A thousand square feet of roof space will yield roughly 600 gallons of water for every inch of rain that falls. Are you letting that asset run down the sewer drain? We need to capture it. This is pure, free, soft water, perfectly balanced for the garden.

LEGALITY FIRST: A CRITICAL CHECK

Do you remember the warning in Chapter 1.4? Before you buy anything, confirm your local zoning laws. Some arid regions restrict rainwater harvesting, believing it impacts downstream water rights. While installing a single 55-gallon barrel is rarely an issue, putting in multiple 300-gallon tanks might require a permit or even be prohibited. Know your jurisdiction. We build resilience within the law, not against it.

THE $49 DOWNSPOUT BYPASS

We are aiming for a simple, low-cost system that requires no pumps for basic irrigation. This system works on gravity alone.

1. The Storage Vessel: Do not compromise on your storage tank. The best options are IBC (Intermediate Bulk Container) totes, which hold 275 to 330 gallons, or 55-gallon drums. If you are using barrels, they must be rated food-grade and, critically, must be opaque (dark green or black). Sunlight penetrating clear or white plastic causes rapid algae growth, fouling your water instantly. Inspect used IBC totes rigorously to ensure they only held food liquids (like syrups or juice concentrates). Never use a container that held chemicals or oil.

2. The Bypass Kit: Purchase a commercial downspout diverter kit. These cost about $40 to $50 and are worth every penny because they manage overflow automatically. You cut a section out of your existing downspout (usually chest high). The diverter replaces that section.

3. Connecting the Tank: Use a flexible hose or simple PVC piping to connect the diverter outlet directly to the inlet on the top of your storage vessel. When the tank is full, the diverter automatically routes the excess water back down the original downspout path, preventing flooding.

FILTER SAFETY CRITICAL: THE FIRST FLUSH

THIS IS THE SINGLE MOST IMPORTANT SAFETY CONCEPT IN RAINWATER harvesting. Your roof is not sterile. It collects dust, pollen, shingle grit, and bird droppings. When a storm begins, the first few gallons of water wash all that debris down the pipe. If this dirty water goes into your tank, it introduces potential pathogens and sediment, ruining the quality of hundreds of gallons of clean water.

YOU MUST INSTALL A FIRST FLUSH DIVERTER

THIS IS ESSENTIALLY A TALL, CAPPED PIECE OF PVC PIPE (OFTEN 3" OR 4" diameter) that intercepts the water before it reaches your main tank. This pipe is designed to hold the first 5–10 gallons of dirty runoff. Once the primary pipe is full, a small floating ball valve seals the pipe shut, forcing the cleaner subsequent rain into your main storage tank.

After the rain stops, the dirty water in the diverter pipe slowly drains out through a tiny weep hole at the bottom, resetting the system for the next storm. This one simple installation keeps your main storage pure.

Additional Filtering: Install fine-mesh screening at two points: where the gutter meets the downspout (to keep out large leaves) and where the diverter hose enters the storage barrel (to prevent mosquitoes from entering the water). Standing, open water is an invitation for pests and disease. Your storage tank must be sealed and secure at all times.

GREYWATER AND RECYCLING: SHOCKING: YOU'RE POURING $500 WORTH OF WATER DOWN THE DRAIN EVERY YEAR! (SIMPLE GREYWATER FILTERS THAT FEED YOUR TREES)

EVERY TIME YOU TURN ON THE WASHING MACHINE, THE SHOWER, OR the bathroom sink, you are creating a valuable resource that most people simply flush away. This used water—greywater—is typically clean enough for irrigation. If you consider that over half of a typical household's water usage comes from these sources, the resource you are literally pouring down the drain is staggering, often amounting to hundreds of dollars in utility costs annually. We need to capture this.

Greywater is strictly defined as wastewater originating from baths, showers, bathroom sinks, and washing machines. For the purposes of simple, safe, small-scale homesteading, we absolutely exclude blackwater, which is sewage from toilets and, critically, wastewater from the kitchen sink (which contains high amounts of grease, food solids, and pathogens). Blackwater requires complex and expensive treatment and should never be used for home irrigation.

Capturing greywater, however, is one of the easiest ways to become water independent.

The core principle is simple: Divert, Filter, Disperse.

The most accessible greywater source for the beginner homesteader is the laundry system—the "laundry-to-landscape" setup. Because laundry water is pressurized, it can be easily diverted through a basic manual valve without requiring a pump.

The system setup is low-tech and effective:

1. The Divert: Install a simple, three-way valve on the discharge line of your washing machine. This allows you to choose between sending the water to the sewer (for very dirty loads or when using chemicals) or diverting it to the landscape system.

2. The Filter: Before the water hits the soil, it must pass through a coarse filter to capture lint, hair, and soap scum. A simple homemade mesh filter bag (made from nylon stocking material or a fine window screen) placed in a bucket is usually sufficient and must be cleaned regularly.

3. The Dispersal: Greywater should never pool on the surface. It must soak into the ground immediately, which is why we utilize a mulch basin. Route the filtered greywater line into a shallow, buried trench (at least 2 inches below the surface) that is filled heavily with wood chips, mulch, or gravel. This deep layer acts as a biological filter, breaking down the organic material in the water before it reaches the plant roots.

NON-NEGOTIABLE GREYWATER SAFETY RULES

For a system to be beneficial, it must be safe and sustainable. Follow these three critical rules:

1. No Storage: Never store greywater. It becomes septic (anaerobic) quickly, producing foul odors and harmful bacteria. The moment it leaves the drain, it must be used for irrigation.

2. Subsurface Irrigation Only: Do not use sprinklers or spray greywater over the top of plants. It should always be dispersed below the mulch level.

3. Use on Non-Edibles (Mostly): Greywater is best used for fruit trees, ornamental shrubs, or perimeter landscaping. Avoid applying it directly to leafy greens, root vegetables, or anything consumed raw. If you use it on food-bearing trees, ensure the water disperses far from the trunk, where the edible parts are not contaminated.

THE SOAP SWAP: WHAT KILLS YOUR SOIL

The biggest threat to your greywater system is standard household soap and detergents.

Most commercial products contain high levels of sodium (salt) and sometimes boron (borax). Sodium is a plant killer; it slowly destroys the soil structure, making it hard, compacted, and impenetrable to water over time. Boron, even in small amounts, is toxic to most plants.

To use greywater successfully, you must switch immediately to garden-safe products:

Look for liquid detergents that are biodegradable, plant-based, and labeled "low-sodium."

Avoid powdered detergents, which often contain harsh fillers.

Avoid all products containing chlorine bleach, borax, peroxygen, or petroleum-based degreasers.

Make the swap to safe soap first. Then, install your simple diverter. In one storm, your roof can give you hundreds of gallons, and in one week, your washing machine can supplement your orchard. This is the definition of reclaiming lost wealth.

—★-★-★—

SIMPLE SOLAR SETUPS: SOLAR POWER MADE STUPID SIMPLE: THE MINIMALIST KIT THAT RUNS YOUR WELL PUMP AND FREEZER WHEN THE LIGHTS GO OUT

THE VAST, INTIMIDATING ROOF-MOUNTED SOLAR ARRAYS YOU SEE ON modern homes are overkill for our immediate goal. We are not looking to power a central air conditioning unit or a home theatre. We are aiming for resilience. We need reliable, emergency power for the absolute essentials: keeping the freezer cold, running a water pump, and maintaining communication.

The minimalist solar setup is built around two simple truths: we only power what we must, and we use equipment designed to be plug-and-play. Forget complex wiring diagrams and expensive utility tie-ins. We are talking about small-scale, DC-based systems that are nearly impossible to wire incorrectly.

The three essential parts of your minimalist kit are: the Collector (Panel), the Storage (Battery/Power Station), and the Critical Load.

STEP 1: DETERMINE YOUR CRITICAL LOAD (THE MATH MADE SIMPLE)

THE BIGGEST BARRIER FOR PEOPLE ENTERING SOLAR IS THE FEAR OF the math. We simplify this by measuring consumption in Watt-Hours (Wh), which is a direct measure of energy use over time.

1. Find the Wattage (W) of the Appliance: Look for the sticker on the back or bottom of the device. A small transfer pump might use 500W. A

chest freezer might only cycle and use 100W on average. A light bulb uses 10W.

2. Determine the Duration (H): How long do you need that item to run in a 24-hour period?

3. Calculate Watt-Hours (Wh): Multiply W x H.

Example: Running the well pump (500W) for 1 hour of total pumping time spread throughout the day, and running a chest freezer (100W average) for 12 hours of cycled run time.

Well Pump: 500W x 1 hour = 500 Wh

Freezer: 100W x 12 hours = 1,200 Wh

Total Daily Need: 1,700 Wh

STEP 2: SIZING THE STORAGE (THE BATTERY)

THE TOTAL NUMBER OF WATT-HOURS CALCULATED (1,700 WH IN our example) is the size of the battery capacity you need.

For simplicity and safety, we strongly recommend using a modern portable power station (often called a "solar generator"). These units combine the battery, the charge controller, and the inverter (to turn the battery's DC power into the AC power needed for most household appliances) into one simple box. They come with output ports for AC wall plugs, USB, and 12V DC.

If your critical load is 1,700 Wh, look for a lithium power station rated for 2,000 Wh or slightly higher. This gives you a crucial buffer and ensures your critical systems can run overnight without sun exposure.

STEP 3: SIZING THE COLLECTOR (THE PANEL)

NOW THAT YOU KNOW THE BATTERY SIZE (E.G., 2,000 WH), YOU NEED a panel strong enough to recharge it in a day of decent sun.

Standard practice assumes about 4 to 5 "Peak Sun Hours" per day in most temperate climates.

To recharge a 2,000 Wh battery in 5 hours, you need a panel that produces 400W (2,000 Wh / 5 hours = 400W).

Look for two easily managed, folding 200W solar panels. These are portable, require no permanent mounting, and plug directly into your power station using standard connectors.

The Minimalist Kit That Works

This minimalist configuration—a 2,000 Wh power station and 400W of folding solar panels—is an investment, but it provides scalable, emergency power that can keep hundreds of dollars of food preserved and provide essential water access during an extended outage.

It is important to run your appliances as efficiently as possible in an emergency. If your well pump is the largest draw, schedule its use: pump water into storage tanks during peak sun hours only, allowing the solar panels to meet the demand directly while still topping off the battery.

This small array runs independently. It is silent. It requires almost no maintenance. It's the simple freedom of knowing your core needs are met, even if the lights are out everywhere else.

BACK-UP HEAT AND COOKING: THE FAIL-SAFE WINTER PROTOCOL: BACKUP HEAT SOURCES THAT REQUIRE ZERO ELECTRICITY (AND HOW TO STORE WOOD SAFELY)

THE GRID IS A FANTASTIC CONVENIENCE UNTIL IT FAILS IN THE DEAD of winter. When the electricity goes, your furnace stops, your water pump quits, and your cooking stove is useless. The fail-safe winter protocol demands reliable backup heat and cooking sources that operate completely independent of the power grid.

The first, non-negotiable step: You must have multiple, tested Carbon Monoxide (CO) detectors placed low to the ground and high on the ceiling in any space where you utilize combustion heat or cooking.

ZERO-ELECTRICITY HEATING SOLUTIONS

FOR SHORT-TERM OUTAGES, PASSIVE DESIGN (CHAPTER 2) HELPS, BUT for multi-day events, you need active heat.

1. Vented Propane/Natural Gas Heaters: These are highly efficient and reliable. Look for models designed for indoor use that feature an ODS (Oxygen Depletion Sensor). While these units are labeled "vent-free," they still produce moisture and use oxygen. They must be operated with a nearby window cracked open to ensure air exchange. Propane is easily stored in standard tanks, making it an excellent, multipurpose emergency fuel source.

2. Small Cast Iron Wood Stove or Heater: If you have a dedicated chimney or vent system, a small, EPA-certified cast iron stove offers the highest

BTU output and longest-term fuel solution. These stoves are excellent heat mass producers and usually have a flat surface suitable for cooking or heating water. In smaller, suburban homes, a portable wood stove is often impractical due to venting requirements, but if you have a hearth or can install a flue safely, wood heat is ultimate security.

3. Kerosene Heaters: An older, reliable option. Kerosene stores well, and the heaters are inexpensive. The downside is the distinct smell and the need to manage ventilation aggressively, as they produce water vapor and combustion by-products.

Critical Note on Location: In a small home, focus on heating one core zone—a bedroom or living space—and closing off the rest of the house. Use thermal blankets and thick curtains to trap the heat you generate.

ZERO-ELECTRICITY COOKING SOLUTIONS

When your electric stove is useless, your options are:

1. Propane Camping Stoves: Standard two-burner camp stoves running off disposable or bulk propane tanks are essential. Use these outdoors if possible. If you must use them indoors, place them near a window with ventilation and run them only as long as necessary.

2. The Rocket Stove: This is the homesteaders' hidden gem. A rocket stove is an extremely simple, portable device that uses the principles of thermal mass and highly controlled airflow to achieve incredible fuel efficiency. It requires only small twigs, sticks, or scrap wood (not large logs) to burn fiercely hot, focusing the heat right onto the bottom of a pot or skillet. They are designed for outdoor use and are perfect for cooking meals with minimal fuel consumption during a crisis.

3. Fire Pit with Grill Grate: The simplest option, utilizing a basic, protected fire pit area in the backyard with a robust metal grate for cooking.

THE FAIL-SAFE FUEL SOURCE: WOOD PROTOCOL

Wood provides the best thermal yield for the long run, but only if prepared correctly. Wet wood is dangerous because it burns inefficiently, produces large amounts of smoke, and leaves creosote (a black, oily residue) that builds up rapidly inside your chimney, leading to chimney fires.

1. Sourcing and Splitting: Source hardwood (oak, maple, ash, hickory) when possible, as it burns longer and hotter than softwood (pine, cedar). Split the wood as soon as it is cut. Splitting exposes more surface area to the air, drastically reducing drying time.

2. Seasoning: Wood needs to dry, or "season," until its moisture content is below 20%. This takes a minimum of six months, and ideally twelve months, for dense hardwoods. Always use wood that has been seasoned; never fresh-cut ("green") wood.

3. Safe Storage Protocols: Wood must be stored off the ground and covered on top, but open on the sides.

Off the Ground: Use pallets, concrete blocks, or stringers to lift the stack at least 6 to 12 inches off the bare earth. This prevents moisture wicking up and keeps insects from burrowing in.

Off the Wall: Stack wood away from sheds, house walls, or fences (ideally 1 to 2 feet away). This ensures maximum airflow and prevents termites or carpenter ants from using your wood pile as a bridge to your structures.

Covered: Place a tarp or roofing material over the top of the stack to shed rain, but leave the sides exposed so the air can flow through and continue the seasoning process.

Stack smart: Always stack the oldest wood closest to the house or stove so it is used first. Follow the rule: Cut today, burn next winter.

CHAPTER FIVE

TURNING SCRAPS INTO GOLD: THE LOW-COST LIVESTOCK SYSTEM THAT PROVIDES $9,251 WORTH OF PROTEIN ANNUALLY.

THIS IS WHERE THE MAGIC TRULY BEGINS.

You have tamed the soil. You have mastered the sun and the rain. Now it is time to invite life onto your land in a more complex, reciprocal way. Integrating livestock—even on a very small scale—is the single most powerful step you can take toward self-sufficiency. This is the moment you close the ecological loop, turning what used to be considered trash or waste into valuable, nutrient-dense gold.

Think back to the last chapter, where we learned to create perfect compost from kitchen scraps. What if you had a system that could convert those scraps even more efficiently, providing high-value protein and ready-made, hot fertilizer for your garden beds, all while performing pest control? That is the role of a low-cost, smart livestock system.

For many, the first steps into animal husbandry feel daunting. There is a deep, instinctual responsibility that comes with being the steward of another creature's life. This is not a task to be taken lightly. When we commit to raising animals—whether for eggs, milk, or meat—we commit to ensuring they live a healthy, safe, and dignified life. We owe them respect. This is the ethical anchor of the modern homestead. If you approach this work with seriousness, compassion, and a commitment to learning, the rewards—in health, flavor, and resilience—will be immeasurable.

The financial return is staggering. The annual $9,251 figure mentioned in the title is not based on selling the produce, but on the conservative replacement cost of the eggs, poultry, and occasional rabbit or pork provided to your family. When you factor in the sheer quality difference—the richness of a true pastured egg, the lean flavor of home-grown rabbit—the value skyrockets further.

But we must be strategic. You are not running a feed lot. You are running a small, efficient food factory. Every creature must pay its way, primarily by converting low-value inputs (weeds, garden waste, kitchen scraps, and low-cost fermented grains) into high-value outputs (eggs, meat, manure). If an animal system requires constant, expensive inputs, it works against the core principle of self-sufficiency.

We will walk through the choices carefully. We will look at creatures designed for small spaces and maximum integration. We will detail coops that defy predators and feed formulas that slash your operating costs by nearly half. This is the step that guarantees protein on your family's table, no matter what happens to the world outside your fence line. Let's start with finding the right companions for your specific space.

CHOOSING YOUR CREATURES: THE R.O.I. OF ANIMALS: WHY CHICKENS ARE THE GATEWAY DRUG AND WHY RABBITS ARE THE SECRET WEAPON OF TINY HOMESTEADS

The decision to bring livestock onto the small homestead marks a definitive shift from gardening to true self-sufficiency. It is often the point of highest anxiety and the highest reward. Animals accelerate the homestead cycle: they turn waste (kitchen scraps, weeds) into usable energy (eggs, meat, milk) and, most critically, into fertilizer, making the entire system regenerative.

But animals are a commitment. They do not take holidays. Therefore, our assessment must focus purely on R.O.I.: Return on Investment. We must measure that R.O.I. in three ways: protein yield, space requirement, and cost of inputs (feed).

The vast majority of homesteaders begin with two key species because they offer the highest R.O.I. per square foot of land: Chickens and Rabbits.

CHICKENS: THE GATEWAY DRUG

Chickens are almost universally recommended as the starting point, and for good reason. They offer immediate, tangible rewards with minimal expertise required.

Space Required: Low to Moderate. You need about 3-4 square feet of coop space per bird and a fenced run (or a secure mobile tractor) for foraging during the day.

Primary Yield: Eggs (daily), Meat (seasonal), Compost/Pest Control.

Feed Cost: Moderate. This is your biggest variable. If you rely solely on commercial feed, your true cost per egg will often equal or exceed store prices. R.O.I. only appears when you supplement aggressively with kitchen scraps, garden waste, and fermented feed (as discussed in Section 5.3).

THE CHICKEN R.O.I. MATRIX

Pro: Daily dividends in the form of eggs. Unbeatable mental reward.

Pro: They are the ultimate compost activators. They turn kitchen scraps and garden beds into fertilized soil quickly.

Con: They require high-level predator defense (foxes, raccoons, hawks). A flimsy fence is a dinner invitation.

Con: Noise. Roosters are prohibited in many suburban zones, and even hens can be surprisingly loud.

Ethical Minimums for Chickens: Hens must have at least 3-4 sq ft of interior, ventilated space, and access to a dust bath (essential for parasite control). They must be secured at night. Never crowd them; stress leads to illness and lower egg production.

RABBITS: THE SECRET WEAPON OF TINY HOMESTEADS

If chickens are the gateway drug, rabbits are the tiny homestead's secret weapon. They are often overlooked because they yield no daily product like an egg, but when measured purely by the efficiency of converting low-cost forage into high-quality protein, they are unrivaled.

Space Required: Very Low. They can be housed in small, stacked hutches or cages, requiring less than 1 square foot of ground space per mature animal, making them perfect for urban settings.

Primary Yield: Meat (rapid cycle), Fertilizer (exceptional quality).

Feed Cost: Low. This is their major advantage. While they need high-protein pellet feed, a large portion of their diet can consist of high-quality hay, garden weeds (dandelion, clover), and specific tree trimmings. This allows for maximum reduction in purchased inputs.

THE RABBIT R.O.I. MATRIX

PRO: SILENT. THIS IS CRITICAL FOR CLOSE-NEIGHBOUR environments.

Pro: Extremely rapid turnover. Meat rabbits (fryers) are ready for harvest in 8-12 weeks, providing continuous protein production throughout the season.

Pro: Manure is 'cold.' Rabbit manure is pelletized and balanced, meaning it can be immediately applied to garden beds without fear of burning the plants (unlike fresh chicken manure). This is free, immediate, premium fertilizer.

Con: They are purely a meat animal. There is no daily product to mitigate the labor cost.

Con: High sensitivity to heat. Rabbits must be kept cool, shaded, and well-ventilated, making summer management crucial.

Ethical Minimums for Rabbits: Housing must be clean and spacious enough for the rabbit to turn and stand fully. Protection from direct sun is mandatory. They must have unlimited access to water and hay. Unlike chickens, rabbits do not naturally thrive in deep bedding; they require wire-bottomed cages (properly sized to protect their feet) that allow waste to fall through.

CHOOSING YOUR FIRST CRITTERS

YOUR CHOICE MUST ALIGN WITH YOUR PRIMARY GOAL.

If your goal is daily food security, pest control, and soil building, start with a small flock of 3-5 heritage breed laying hens.

If your goal is maximizing meat protein yield and high-quality, immediate fertilizer on a zero-lot line, the rabbit system offers the fastest, quietest path to independence.

Regardless of the animal, remember this fundamental rule: your time is the most expensive input. Select breeds known for hardiness and temperament, and always prioritize secure, well-ventilated housing over anything else. A small investment in high-quality hardware cloth now will save you from a heart-breaking predator attack later.

——★-★-★——

COOP DESIGN: EXPOSED: THE 3 FATAL DESIGN FLAWS THAT INVITE FOXES (THE $7 'HARDWARE CLOTH BARRIER' NO PREDATOR CAN BREACH)

The most tragic morning on the homestead is the one where you discover a predator breached your defences. Predators are relentless, intelligent, and motivated by instinct. Your coop is not just a home for your flock; it is a fortress that must repel calculated attacks.

There are three common, fatal design flaws that practically invite foxes, raccoons, and weasels. Avoid them, and your coop will be secure.

FLAW 1: RELYING ON CHICKEN WIRE

Chicken wire (or poultry netting) is designed to keep chickens in. It is not designed to keep predators out. The thin, flexible wire can be ripped by a determined raccoon, pushed open by a fox, and easily chewed through by smaller rodents like rats.

The mandatory upgrade is 1/2-inch galvanized hardware cloth. This material is rigid, welded, and virtually impossible for common backyard predators to chew or tear. It is the $7 difference per roll that ensures the safety of your entire investment. All windows, vent openings, and the perimeter of the run must be covered with hardware cloth secured with heavy-duty staples or screws and washers.

FLAW 2: STOPPING THE BARRIER AT GROUND LEVEL

Foxes and dogs don't climb fences often; they tunnel under them. A solid wall that stops at the dirt is an invitation to dig.

The solution is the 'Hardware Cloth Apron.'

This barrier prevents digging attempts directly adjacent to the coop wall. When installing your run enclosure:

1. Dig a shallow trench, about 2 inches deep, running along the entire perimeter of the coop and run.

2. Extend the 1/2-inch hardware cloth vertically up the wall, and then lay it horizontally outward into the trench for at least 12 to 18 inches.

3. Secure the cloth firmly to the base of the wall.

4. Backfill the trench with dirt, sod, or heavy gravel.

When a predator attempts to dig near the wall, they immediately hit the horizontal apron and become discouraged. They rarely attempt to dig 18 inches away from the wall. This simple addition eliminates 90% of burrowing attacks.

FLAW 3: USING SIMPLE LATCHES

Raccoons have incredible dexterity. Any simple hook-and-eye or swivel latch can be opened by a curious raccoon within minutes.

Security Step-Up: Always use heavy-duty barrel bolts or padlocks. For ultimate nocturnal security, use spring-loaded clips or carabiners on all doors and access points. Double-latching is mandatory—one latch at the top, one at the bottom. The coop must be sealed airtight every evening.

OPTIMAL INTERNAL DESIGN: VENTILATION AND CLEANABILITY

A SECURE COOP MUST ALSO BE A HEALTHY ONE. THE TWO MOST overlooked elements in coop design are ventilation and ease of cleaning.

Ventilation is life-saving. Chickens generate huge amounts of water vapor and ammonia, especially overnight. Poor ventilation leads to respiratory illness (pasting up) and, ironically, frostbite in winter, as moist air freezes rapidly.

1. High Vents: Build fixed vents near the roof line, always above the roosting bars. These vents allow warm, humid, ammonia-filled air to exit. These must be covered with hardware cloth.

2. Low Vents: Provide low, adjustable vents near the floor to bring in fresh, cool air. In extreme cold, these can be partially closed, but the high vents must always remain open. Never position vents directly at roost level, as drafts cause illness.

CLEANING FOR HEALTH: THE DROPPINGS BOARD

CLEANING CHICKEN MANURE IS TEDIOUS, BUT CRITICAL FOR CONtrolling pests and disease. The easiest way to manage waste is with a removable droppings board positioned directly under the roosting bars.

Chickens sleep and defecate directly under the roosts. By installing a plywood or vinyl-covered board (or a tray) a few inches beneath the roosts, you capture the majority of the waste. This allows for a quick daily scrape-off into the compost pile, rather than cleaning the entire floor area weekly. This dramatically reduces ammonia buildup and keeps your flock healthy.

The final secure coop checklist is simple:

1. Hardware cloth on all openings (including the floor if building a mobile run).

2. Hardware cloth apron buried horizontally 18 inches around the perimeter.

3. Dual, predator-proof locking mechanisms on all doors.

4. Adequate high and low ventilation.

5. Droppings board for easy daily clean-up.

—★—★—★—

DAILY CARE AND FEED MANAGEMENT: THE 'FERMENTED FEED FORMULA': CUTTING YOUR MONTHLY ANIMAL FEED BILL BY 40% USING KITCHEN SCRAPS AND WHEY

THE SINGLE LARGEST ONGOING EXPENSE IN A SMALL LIVESTOCK OPeration is feed. If you are buying premium commercial feed, your monthly budget can evaporate quickly. The goal is to maximize the value of every dollar spent on commercial feed while aggressively utilizing free, high-nutrient supplements found right on your property. This approach can reliably cut your dependency on commercial sources by 30 to 40 percent.

The cornerstone of this feed economy is transformation. Do not feed your animals dry grain when you can feed them living, expanded, fermented grain.

THE 'FERMENTED FEED FORMULA'

FERMENTING YOUR ANIMAL FEED ACHIEVES TWO THINGS: IT RADIcally increases nutrient availability, and it expands the physical volume of the feed, meaning your animals feel satiated on less. This method is essential for chickens, ducks, and even rabbits (though care must be taken with the type of grain offered to rabbits).

1. The Base Grain: Use your standard pellet or whole grain feed. If using whole grain, roll or slightly crack the kernels to aid in fermentation.

2. The Liquid: Water is fine, but for superior probiotic power, use whey (a by-product of cheesemaking—see Chapter 7) or leftover kombucha brew. These introduce diverse beneficial bacteria.

3. The Method: Place the feed in a clean, non-metallic container (a five-gallon food-grade bucket is ideal). Cover the grain entirely with liquid, ensuring an extra inch of liquid on top. Cover the container with a cloth (not an airtight lid, as fermentation produces gas).

4. The Time: Allow it to sit at room temperature (65 to 75 degrees F) for two to four days. You will see bubbles and smell a distinct, slightly sour, yeasty aroma (like sourdough). This is ready.

5. Feeding: Feed the animals the entire contents (liquid and solids). Because the feed has expanded, you will find you use significantly less volume each day. Never let the feed go moldy or smell rancid; if it does, discard it and start fresh.

THE KITCHEN SCRAPS AND WEED SUPPLEMENTATION

YOUR GARDEN AND KITCHEN ARE NOT JUST SOURCES OF HUMAN food; they are recycling centres. Never let valuable nutrients go into a landfill.

Safe Supplements:

Cooked rice or pasta, bread ends (in moderation), safe garden weeds (dandelion, chickweed, clover), vegetable peelings (carrot, beet, pumpkin), fruit scraps (apple cores, berries), and leftover cottage cheese or yogurt. These supplements should make up 20 to 30 percent of the total diet.

Unsafe Supplements:

Avoid raw potato peels, green tomatoes, avocado pits and skins, dried beans, citrus fruits (they cause digestion issues in poultry), and anything moldy or excessively salty. Never feed these to your animals.

Maximized Foraging: For poultry, utilize the Chicken Tractor model (moving a bottomless cage over spent garden beds or lawn areas). The animals scratch, weed, and fertilize simultaneously, effectively producing some of their own feed through insect and grass consumption, reducing your feed bill while improving the land.

DAILY CARE AND OBSERVATION: THE BEST MEDICINE

Preventative care is always cheaper than treating a sick animal. Your primary job as a steward is observation. Spend five minutes near your flock or colony every day, not just to feed them, but to look at them.

1. Water Management: Clean water is critical. Stagnant, dirty water is the fastest way to introduce bacteria and parasitic diseases. Scrub out water vessels daily. If the water looks murky, dump it.

2. Visual Inspection (The Early Warning Signs):

For Chickens: Look at the comb and wattles. They should be bright red and firm. Pale, droopy, or blue combs suggest illness or distress. Check their eyes—they should be clear and bright. Watch for listlessness or an animal separating itself from the flock.

For Rabbits: Check the eyes and nose for discharge. Look for wetness around the dewlap or mouth (which can signal teeth problems). Check the hutch for diarrhea, which is often a fatal sign in rabbits.

3. Manure Check: A healthy animal produces healthy manure. For chickens, it should be firm with a white cap (urates). Runny, greenish, or

bloody manure indicates a problem with parasites or diet, requiring immediate attention.

4. Housing Cleanliness: Cleanliness prevents pests and disease. Remove soiled bedding from coops and hutches daily or every other day, and replenish with fresh straw or shavings. A clean environment means less stress and better immunity, reducing the need for expensive veterinary interventions.

—★–★–★—

PROCESSING AND ETHICAL HARVEST: THE HUMANE HARVEST PROTOCOL: CLEAR, RESPECTFUL STEPS FOR HOME PROCESSING (THE RESPONSIBILITY 99% OF CONSUMERS AVOID)

This is the critical juncture where the philosophy of self-sufficiency meets raw reality. If we choose to integrate animals into our homestead for sustenance, we accept the most profound responsibility: we must be willing and able to participate fully in the cycle of life and death. We cannot outsource the difficult parts while claiming the moral high ground of ethical consumption. The humane harvest is the final act of stewardship, and it demands our full respect and preparation.

The quality of the animal's life is paramount, but the quality of its passing is non-negotiable. It must be swift, quiet, and decisive.

PREPARATION: THE FOUNDATION OF RESPECT

Before the day of harvest, every single detail must be accounted for. Hesitation, fumbling, or realizing you lack a necessary tool is what leads to suffering.

1. Know the Local Law: Confirm any local ordinances regarding home processing. Some jurisdictions have rules concerning visible butchering or disposal of offal.

2. Tools and Setup: Everything must be clean, sharp, and accessible. For poultry, this means a clean scalding pot (kept at the correct, consistent temperature), clean processing cones, and razor-sharp knives. For rabbits, this means a clean, dedicated processing station and a method for swift dispatch.

3. Stress Reduction: The animal must be calm immediately prior to dispatch. Stress releases hormones that severely degrade meat quality. Handle the creature gently, keep the environment quiet, and never dispatch an animal within sight or sound of its living companions.

THE HUMANE HARVEST PROTOCOL: DECISIVENESS IS KINDNESS

THE SINGULAR GOAL OF THE DISPATCH IS IMMEDIATE, IRREVERSIBLE unconsciousness. Any method chosen—be it cervical dislocation for small birds, or the use of a captive bolt or sharp blow for larger stock—must be mastered beforehand and performed without second-guessing.

1. The Dispatch: Employ the chosen method with absolute decisiveness. The instant application of force must render the animal completely insensible.

2. Exsanguination (Bleeding): Immediately following the dispatch, the animal must be bled out completely. This is the physiological action that ensures death and prepares the meat for safe consumption. For poultry, this involves a swift cut to the major veins in the neck. Collect the blood in a designated area or vessel.

3. Prompt Processing: Once the animal is fully deceased, the cleaning and processing must begin immediately. Heat transfers quickly to muscle

tissue, and speed is required to maintain cleanliness and prevent bacterial growth.

A NOTE ON SOLITUDE AND GRATITUDE

This is not a task for an audience. It requires solemn focus. When you are done, take a quiet moment. Recognize the sacrifice made. This isn't just a piece of protein; it is the culmination of your labor, the gift of the creature's life, and the bedrock of your family's security.

This responsibility—the willingness to face the full cost of your food—is the deepest expression of self-sufficiency and the highest form of respect for the food chain. Never take it lightly.

CHAPTER SIX

—★-★-★—

THE HARVEST DEFENSE STRATEGY: HOW TO CAPTURE 90% OF YOUR GARDEN'S VALUE BEFORE IT SPOILS (PRESERVATION MASTERY).

THERE IS A MOMENT IN THE LATE SUMMER, OR MAYBE THE EARLY AUtumn, when the garden shifts from giving to demanding. It's a beautiful chaos. The tomatoes turn red faster than you can pick them, the beans swell overnight, and the squash threatens to overrun the path. This is the payoff, the grand finale, the moment your entire season's worth of work comes due.

This moment also represents the highest vulnerability of the entire self-sufficiency journey.

You spent months planning the soil, weeding the rows, fighting the pests. But if you fail at preservation, you lose the war. All that energy, all that water, all that sun—it simply reverts back to soil, or worse, compost, within a week or two. You must transition instantly from the role of Gardner to the role of Collector and Defender.

Preservation is the Harvest Defense Strategy. It is the critical barrier between abundance and scarcity.

Think of it this way: When you save a jar of tomatoes in July, you aren't just saving food; you are banking labor and sunlight. That jar is a currency that doesn't suffer inflation. It is food security sealed under a lid.

We have forgotten how vital these heritage skills are. For generations, the difference between life and starvation through winter was dictated by how much was successfully canned, dried, or stored underground. We traded that profound, life-giving knowledge for the convenience of the year-round, sterile grocery aisle. Now, we reclaim it.

Reclaiming preservation is perhaps the most spiritually rewarding part of the homesteading cycle. There is deep satisfaction in looking at a pantry shelf lined with jewel-toned jars—a visual ledger of your victory over the impending winter. Every jar is a small act of defiance against a system that wants you dependent.

But you have to move fast. The trick to preservation mastery is having your equipment and plan ready before the flood arrives. You need to know exactly which method is best for which crop: canning the high-acid tomatoes, drying the herbs, cellaring the sturdy root vegetables.

We aren't just aiming to save a little bit. We are building a fortress of food. Our goal is simple and aggressive: capture 90% of your garden's value before it spoils. That means mastering the safety protocols of canning, understanding the science of moisture control in drying, and controlling the temperature and humidity for long-term storage.

Let's turn that overwhelming harvest into lasting, reliable wealth.

—★-★-★—

THE ART OF CANNING: THE BOTULISM BOOGEYMEN: THE SINGLE CRITICAL STEP THAT MAKES PRESSURE CANNING 100% SAFE (WHEN TO USE WATER VS. PRESSURE)

The thought of canning often conjures images of bubbling jars and, worse, the spectre of botulism. We need to strip that fear away. Canning is a scientific process. When you follow the rules, it is 100% safe. Failure to follow the rules is what causes danger.

The single, critical step that makes pressure canning safe is simple: You must get the food hot enough, long enough, to kill the spores of Clostridium botulinum. These spores are the root of the botulism boogeymen, and they thrive in low-acid, airless environments.

Water boils at 212 degrees Fahrenheit (100°C). This is not hot enough to kill those spores.

This is the distinction that divides all canning: acidity.

WATER BATH CANNING: FOR HIGH-ACID FOODS

If the food is acidic enough, the botulism spores cannot grow or produce toxins, even if they survive the boiling process.

Use the water bath method for:

Jams and jellies.

Fruits.

Pickled vegetables (cucumbers, carrots, etc.).

Salsa.

Tomatoes (You must often add acid—bottled lemon juice or citric acid —to ensure modern varieties are safe).

The equipment is simple: a large pot with a lid, a rack to keep jars off the bottom, tongs, and canning jars. The jars are submerged in boiling water and processed for a specified time.

PRESSURE CANNING: FOR LOW-ACID FOODS

Low-acid foods—anything with a pH above 4.6—must be processed at temperatures higher than boiling water. The only way to achieve these temperatures at home is with pressure. The pressure holds the heat above 212°F, typically reaching 240°F, which effectively destroys the botulism spores.

This is the single critical step: Always use a pressure canner for low-acid foods. There are zero exceptions. Do not try to water bath can these items.

Use the pressure canning method for:

Vegetables (corn, potatoes, green beans, carrots).

Meats (chicken, beef, fish).

Soups and stocks.

Chili and baked beans (non-pickled).

The equipment needed is a reliable pressure canner (not a pressure cooker, which is different) and standard canning jars. The canner must have a gauge—either a weighted gauge or a dial gauge—to monitor the pressure and temperature accurately.

Never rely on folklore or old, unverified recipes. Use processing times and procedures published by major university extension offices or the National Center for Home Food Preservation (NCHFP). They have done the science. Your job is only to follow the map exactly. When in doubt, pressure can. Safety is non-negotiable.

FREEZING AND DEHYDRATING: THE BLANCHING BYPASS: HOW TO FREEZE YOUR PEAS AND BEANS WITHOUT LOSING FLAVOR OR NUTRIENTS (THE QUICK PRESERVATION TRICK)

The harvest comes at full speed, and if you have 20 pounds of green beans hitting their peak on the same day, you need preservation methods that are fast, efficient, and reliable. Freezing and dehydrating are the speed demons of preservation, allowing you to capture peak nutritional value in a fraction of the time required for canning.

The primary enemy of frozen vegetables is enzyme activity. Even at freezing temperatures, naturally occurring enzymes within the vegetable continue to degrade its color, flavor, and nutrient content over a few months. This process leads to freezer burn and limp, flavorless vegetables when thawed.

The solution is blanching. Blanching is the quick heat treatment that neutralizes these enzymes, allowing the vegetable to maintain its vibrant color and fresh taste for up to a year in the freezer.

THE STANDARD BLANCHING PROTOCOL

1. Prepare: Clean and chop your vegetables (e.g., snap beans, broccoli, carrots) into serving sizes.

2. Heat Shock: Drop the vegetables into vigorously boiling water. Timing is critical and short. Most vegetables require 1.5 to 4 minutes. Start timing only after the water returns to a full boil.

3. Cold Shock: Immediately scoop the blanched vegetables out and plunge them into an ice-water bath for the exact same amount of time they boiled. This stops the cooking process instantly.

4. Dry and Freeze: Drain the vegetables completely. For best results, spread them in a single layer on a sheet pan (this is called "flash freezing") until solid. Once frozen, transfer them to airtight freezer bags or vacuum-sealed containers, expelling as much air as possible. Label and date immediately.

The Blanching Bypass

While blanching is essential for most hardy vegetables like beans, peas, and cauliflower, there are preservation opportunities where we can bypass the process entirely for speed.

1. High-Water Fruits: Berries, mangoes, and chopped peaches do not require blanching. Wash them, dry them thoroughly, and flash freeze them on trays before bagging.

2. Soup and Sauce Bases: If the vegetable is destined for a long-cooked soup, stew, or casserole, some homesteaders choose to skip blanching on items like corn kernels or zucchini. The future cooking process will destroy the enzymes anyway. Note: This compromises flavor and texture for stand-alone side dishes but saves significant time when bulk processing.

3. Herbs: Herbs like basil, cilantro, and parsley should never be blanched, as they turn to mush. Instead, chop them finely and freeze them in ice cube trays with a small amount of water or olive oil.

DEHYDRATING: SHRINKING THE HARVEST

Dehydration removes water content, stopping decay and shrinking the volume of the harvest dramatically. It is perfect for herbs, mushrooms, sliced fruits (apples, bananas), and making vegetable powders.

The Dehydrator Advantage

A dedicated dehydrator provides consistent, low heat (typically 125°F to 135°F) and constant airflow, which are the keys to successful drying. Simply slice foods thinly and uniformly, place them on the trays, and let the machine run until the items are brittle (like chips).

The Low-Cost Oven Method

If a dehydrator is not yet in the budget, your oven can serve as a suitable backup, though it is less energy efficient.

1. Set the Temperature: Set your oven to its lowest possible temperature, ideally 140°F to 170°F. Anything higher will cook or burn the food instead of dehydrating it.

2. Ventilation is Key: Prop the oven door open slightly—use a wooden spoon or a small piece of heat-safe foil. This allows the evaporated moisture to escape. If the moisture cannot escape, the oven essentially steams the food.

3. Preparation: Slice all items uniformly thin (1/8 to 1/4 inch is ideal). Place them on baking sheets lined with parchment paper or cooling racks to allow air circulation.

4. Rotate and Monitor: Drying time will vary dramatically (often 4 to 12 hours). Rotate the racks every hour to ensure even drying.

Final Storage Check: Whether frozen or dehydrated, ensure the preserved food is bone dry (for dehydrated) or solidly frozen (for blanched goods) before storing. Condensation and residual moisture are the leading causes of preservation failure. Store dehydrated goods in completely airtight jars, away from direct light, to prevent reabsorption of humidity.

ROOT CELLARS AND COOL STORAGE: DISCOVERED: HOW TO BUILD A $50 DIY ROOT CELLAR IN YOUR BACK YARD THAT KEEPS POTATOES FRESH FOR 9 MONTHS

THERE IS NO GREATER SATISFACTION THAN OPENING YOUR OWN pantry of stored treasures, months after the ground has frozen. This is preservation without electricity, leveraging the earth's natural thermostat to sustain your harvest. When done correctly, cool storage allows you to keep hard crops—potatoes, apples, carrots, cabbage—fresh for six to nine months, capturing the full value of your garden before it spoils.

The core physics of a root cellar is simple: replicate the stable, cold environment of a deep cave. We need four things: consistent cool temperatures (32-40°F), extremely high humidity (85-95%), total darkness, and controlled ventilation.

UTILIZING EXISTING SPACE: THE BASEMENT HACK

THE EASIEST ROUTE, IF YOU HAVE ONE, IS TO UTILIZE AN UNHEATED corner of your basement or crawl space. These spaces often hover around 50°F, which is close enough.

1. Isolation: Wall off a small 4x4 or 4x6 foot section using inexpensive rigid foam insulation board or thick plastic sheeting stapled to studs. This creates a thermal break from the heated areas of the house.

2. COOLING AND VENTING: INSTALL A SMALL INTAKE VENT NEAR THE FLOOR, LEADING TO THE OUTSIDE. INSTALL AN EXHAUST VENT NEAR THE CEILING, ALSO LEADING OUTSIDE. USE SMALL, SCREENED OPENINGS WITH DAMPERS YOU CAN CLOSE DURING EXTREME COLD. THIS PASSIVE AIRFLOW IS ESSENTIAL FOR CIRCULATING FRESH AIR AND PREVENTING THE BUILDUP OF ETHYLENE GAS (WHICH CAUSES PREMATURE RIPENING AND SPROUTING).

THE $50 BACKYARD PIT CELLAR

If a basement is not an option, you can build a highly effective, low-cost pit cellar right in your backyard.

1. Dig the Pit: Dig a hole about 4 feet deep and 4 feet in diameter. The depth is crucial because the earth below the frost line maintains a stable temperature year-round.

2. Structure: Line the pit with wood pallets, straw bales, or old cinder blocks for stability. You want to avoid direct contact between your vegetables and the native soil, which can harbor pests.

3. Storage Vessels: Place large plastic storage bins, wooden crates, or heavy-duty garbage cans (with lids) inside the pit. These will hold your produce.

4. Insulation and Lid: Cover the top of the pit with a sturdy, insulated lid. Use thick plywood topped with rigid foam insulation, or even a layer of straw and then plastic sheeting. Seal the lid well, leaving only a small, screened ventilation pipe (a simple PVC drain pipe works) running up out of the lid to provide necessary air exchange.

THE HUMIDITY MASTERY PROTOCOL

THIS IS THE MAKE-OR-BREAK ELEMENT. WITHOUT HIGH HUMIDITY, potatoes shrivel and carrots become rubbery. If the humidity is too high and the ventilation is poor, everything mildews.

To achieve that critical 85-95% relative humidity:

Water Pans: In a concrete basement cellar, place shallow metal pans of water on the floor. Evaporation will raise the ambient moisture.

The Earthen Advantage: If you have an earthen floor (in a crawlspace or pit cellar), simply dampening the floor soil with a hose once a week will often maintain perfect humidity.

The Sand/Peat Moss Layer: The golden trick for storing root vegetables like carrots, beets, and parsnips is to layer them with a moist medium. In your crates or bins, alternate layers of vegetables with slightly damp sand, peat moss, or sawdust. This keeps moisture locked around the individual root while preventing the vegetables from touching each other, which stops the spread of rot. The medium should feel damp, not wet, like a sponge you have squeezed out completely.

STORAGE PREPARATIONS FOR KEY CROPS

POTATOES: CURE THEM FOR TWO WEEKS IN A DARK, HUMID SPACE AT 50-60°F before storing them in the root cellar. Then, store them in burlap sacks or open crates in total darkness at cellar temperatures. Keep them as far away from apples as possible, as the ethylene gas from apples encourages sprouting.

Carrots, Beets, Parsnips: Trim all greens completely to prevent them from drawing moisture out of the root. Layer them in damp sand or sawdust, as described above.

Apples: Harvest only undamaged apples. Wrap each apple individually in newspaper or tissue paper. This prevents a single spoiled apple from releasing mold spores that ruin the rest of the barrel. Store apples on high shelves, separate from the potatoes and cabbage, because they are heavy ethylene producers.

—★-★-★—

CURING AND SMOKING MEATS: THE HALBERT METHOD OF CURING: TURNING CHEAP CUTS OF PORK INTO ARTISAN BACON USING ONLY SALT AND TIME (THE HERITAGE SKILL)

THE SKILL OF SAVING MEAT IS ONE OF THE OLDEST HERITAGE SKILLS we possess. Before refrigeration, a farmer's survival over winter depended entirely on his ability to use salt and smoke to transform perishable meat into stable, storable food. We have largely forgotten this skill, but it is easy to reclaim, and the results—like a slab of homemade bacon—are infinitely superior to anything you can purchase.

The simplest place to begin is with dry curing bacon. We call this the Halbert Method because it relies on the principle of using salt, time, and cool temperatures, which is how our great-grandparents managed their winter stores.

The key to curing is salt. Salt inhibits the growth of bacteria by drawing moisture out of the meat, lowering the overall water activity to a point where spoilage organisms cannot thrive.

STARTING WITH BACON: DRY CURING

YOU WILL NEED A PORK BELLY, WHICH IS OFTEN ONE OF THE LEAST expensive cuts available. Choose a piece that is roughly uniform in thickness.

1. Prepare the Cure Mix: Use non-iodized salt (Kosher or sea salt works best). While traditional methods often use just salt, modern recipes usually incorporate a small amount of sugar (for flavor balance and browning) and, critically, pink curing salt #1 (sodium nitrite) for safety. For true preservation, especially against the deadly threat of botulism spores, the nitrite is a modern necessity. However, for a quick cure focusing on the heritage skill, a heavy salt cure followed by immediate freezing or prompt consumption works, but for safety and long-term storage, always include the proper proportion of curing salt #1. A common ratio is one tablespoon of cure mix (salt, sugar, nitrite) per pound of meat.

2. Rub and Rest: Thoroughly rub the entire surface of the pork belly with the cure mix. Ensure every surface, including the edges, is coated. Place the meat into a large plastic zip-top bag or a clean, non-reactive container.

3. The Cure Time: This is where time does the work. Place the meat in the refrigerator (critical temperature control: always below 40°F). Cure for 7 to 10 days, depending on the thickness of the meat. Flip the bag or container daily to redistribute the cure and ensure even penetration. When the meat feels firm to the touch, it is fully cured.

4. Rinse and Dry: Remove the bacon and rinse it thoroughly under cold water to remove all excess salt and cure mix. Pat it completely dry. Now, place the cured belly uncovered back in the refrigerator for 12 to 24 hours. This allows a sticky film called a pellicle to form on the surface. The pellicle is essential—it helps the smoke adhere to the meat.

SMOKING FOR FLAVOR

While the salt cure is the preservation step, smoking adds depth of flavor and another layer of surface preservation.

For the small-scale homesteader, hot smoking is the easiest and safest method. Hot smoking cooks the meat while simultaneously flavoring it.

To hot smoke your cured bacon, you need a smoker capable of maintaining a consistent temperature between 180°F and 225°F. Use fruitwoods (apple, cherry) or nutwoods (hickory, pecan) for the best flavor. Smoke the meat until the internal temperature reaches 150°F. Use a reliable meat thermometer. This step kills surface bacteria and locks in flavor. Once done, let it cool completely before slicing.

EXTENDING THE SKILL TO FISH AND POULTRY

The same principles of salt and temperature apply to other meats. Fish, like small trout or fillets of salmon, can be brined in a heavy salt solution for several hours and then hot smoked. Because fish has less connective tissue, the curing time is significantly shorter, sometimes just a few hours.

Poultry—often chicken or duck breast—can also be cured and smoked. The curing time will be brief (12 to 24 hours) due to the thinness of the cut, but the temperature control is absolute: the meat must remain below 40°F during the cure and must hit the necessary safe internal temperature during the hot smoke.

Never forget: Curing is a science of control. Measure your salt precisely, and monitor your temperatures religiously. These two variables are the difference between a lifetime supply of artisan meat and a dangerous mistake. Respect the salt, and you will honor the skill.

CHAPTER SEVEN

KITCHEN ALCHEMY: MASTERING THE 'TRANSFORMATION TAX' (CONVERTING $1 OF RAW MILK INTO $15 WORTH OF ARTISAN CHEESE).

YOU'VE PUT IN THE LABOR. THE SUN AND RAIN COOPERATED. YOU'VE harvested the bounty—crates of tomatoes, buckets of milk, baskets of cucumbers. This is the moment most people stop. They freeze what they can, can a few jars, and then watch the rest spoil or rot. That is money, time, and life energy wasted.

We aren't going to let that happen.

Your kitchen is not just a place to warm up leftovers. It is a laboratory. It is the single most valuable piece of real estate on your entire homestead, regardless of size. This is where true alchemy happens: the "Transformation Tax." It's the skill-based uplift that takes a simple quart of raw milk (worth perhaps $1.50) and, with a few hours of focused attention and patience, turns it into a wheel of fresh mozzarella or a batch of aged cheddar worth ten times that.

Think about your surplus. A mountain of peppers can be dried, toasted, and ground into priceless paprika. Excess milk becomes cream, butter, and cheese. Waste vegetables turn into immune-boosting ferments. We are not just preserving food; we are creating concentrated wealth and flavor that cannot be replicated at the store.

These skills are heritage skills. They were known by every great-grandmother and, tragically, lost in a single generation of reliance on industrial supply chains. We are taking them back. Learning these processes is not difficult, but it requires patience, absolute cleanliness, and confidence.

Let's step into the lab. We will focus on four core areas of transformation: harnessing the power of simple flour and water, turning milk into masterpieces, mastering the living world of fermentation, and crafting our own medicinal remedies straight from the garden. Mastering these four skills ensures that your pantry is never empty, and your health is never compromised. We are trading passive consumption for active creation. And the results taste infinitely better.

—★–★–★—

SCRATCH BAKING: THE 5-MINUTE STARTER SECRET: BAKING WORLD-CLASS SOURDOUGH BREAD WHEN THE GROCERY SHELVES ARE EMPTY

If bread is life, then sourdough is the breath of true self-sufficiency. You do not need yeast packets, which spoil or run out. All you need is flour, water, and patience. The 5-Minute Starter Secret is not about magically creating the culture overnight; it's about establishing the culture and then effortlessly maintaining it for decades.

Your starter is a living colony of wild yeast and bacteria (lactobacilli). It is robust, resilient, and utterly dependent on you for its food.

THE STARTER MAINTENANCE PROTOCOL

1. Creation (Initial Phase): Mix 50 grams of whole wheat flour (organic is best) with 50 grams of non-chlorinated water in a clean jar.

Stir. Cover loosely. Let it sit on the counter for 24 hours. The next day, discard half, and feed it another 50g flour and 50g water. Repeat this process every 24 hours. After 5 to 10 days, the starter will consistently double in size within four to six hours after feeding. You have success.

2. The Fridge Protocol (The Secret): If you bake once a week or less, you do not need to feed your starter every day. This prevents massive waste and burnout. Once your starter is strong, feed it, let it rise slightly (about an hour), then screw on the lid and place it in the refrigerator. The cold puts the microbes into a slow, dormant sleep.

3. The 5-Minute Revival: When you want to bake, pull the starter out. Take out about 50 grams of the refrigerated starter. Feed this active portion 100 grams of flour and 100 grams of water (a 1:2:2 ratio works wonders for a quick revival). Let it sit on the counter until it is bubbly and at its peak—usually 4 to 8 hours. The process of pulling, measuring, and feeding takes less than five minutes. Use the peak portion for your dough and immediately put the remaining mother culture back into the fridge. That's all there is to it.

THE HIGH-ALTITUDE SECRET: MILLING YOUR OWN GRAIN

You can bake decent bread with store-bought flour, but you will never achieve world-class bread without understanding the quality of the raw ingredient. Store-bought white flour is often stripped of its most nutritious and flavorful components (the germ and the bran) and has been sitting on shelves for weeks, if not months. This allows it to oxidize, losing flavor and nutritional value.

The key to elevating your baking lies in milling your own grain, even in small batches.

A small tabletop grain mill (electric or manual) is one of the most powerful, overlooked assets on a self-sufficient homestead. It takes seconds to mill the exact amount of flour you need for a recipe.

The benefits are immediate:

1. Maximized Flavor: Freshly milled flour releases aromatic oils and flavor compounds that give the bread a deep, complex, nutty profile that industrial flour cannot match.

2. Complete Nutrition: You retain the germ, which contains essential fats and B vitamins. You are baking with the whole package.

3. Starter Vigor: Your sourdough culture thrives on the complex sugars found in fresh, whole grains. Feeding your starter with fresh-milled grain makes it stronger, more active, and more reliable, especially if you are working with grains you grew yourself.

By controlling the flour quality and mastering the simple maintenance of a sourdough culture, you gain permanent access to the foundational food source of human civilization. The shelves can be bare, but your counter will always yield a loaf of warm, resilient perfection.

MAKING CHEESE AND DAIRY: THE 'ACID TEST': MAKING CREAMY, FRESH MOZZARELLA AND RICOTTA IN YOUR KITCHEN IN UNDER 30 MINUTES

THIS IS WHERE SIMPLE INGREDIENTS UNDERGO A PROFOUND, ALMOST unbelievable transformation. You are about to convert liquid milk into solid, delicious food that costs pennies on the dollar compared to the imported artisan versions. This is the ultimate "transformation tax" payoff.

Most people think cheesemaking requires exotic cultures, expensive equipment, and a week of waiting. Not true. We can utilize the simplest tool—acid—to create incredible fresh cheese in less time than it takes to order takeout.

THE SINGLE CRITICAL RULE: MILK MATTERS

Before you start, check your label. You must use milk that is pasteurized but not Ultra-Pasteurized (UHT). Ultra-Pasteurization heats the milk so intensely that it chemically alters the proteins, preventing them from forming strong curds when acid is introduced. If your milk is UHT, this won't work. Look for whole milk (preferably local or non-homogenized, if possible, but standard pasteurized whole milk works fine).

QUICK RICOTTA: THE 15-MINUTE KITCHEN MIRACLE

Ricotta literally means "re-cooked," as it was traditionally made from the whey left over from harder cheeses. But the fastest and most foolproof version uses whole milk and acid to create the lightest, creamiest cheese imaginable.

What You Need:

1 Gallon whole milk (non-UHT)

1/2 Cup White Vinegar (distilled) or 1/2 Cup Lemon Juice

1 Large non-reactive pot (stainless steel)

A fine-mesh sieve or colander lined with cheesecloth

Thermometer (optional, but highly recommended)

The Steps (The Acid Test):

1. Heat the Milk: Pour the milk into the pot. Heat slowly over medium-low heat, stirring constantly to prevent scorching. We are aiming for 190°F (88°C). If you don't have a thermometer, heat until the milk is steaming heavily and just about to bubble aggressively. Do not let it boil.

2. Add the Acid: Remove the pot from the heat source. Gently stir in the 1/2 cup of vinegar or lemon juice.

3. Wait for the Magic: Stop stirring. Cover the pot and let it sit undisturbed for 10 minutes. This is the moment of transformation. The acid will immediately react with the heated milk proteins, causing them to coagulate and separate into white, fluffy curds and translucent, yellowish whey.

4. Strain: Carefully ladle the curds into your cheesecloth-lined sieve or colander. Allow the whey to drain completely (about 5 minutes for a fresh, wet ricotta).

5. Finish: Transfer the drained curds to a bowl. Season with salt to taste. Serve immediately, or store in the refrigerator. Use the leftover whey for baking or feeding your animals (Chapter 5).

FRESH MOZZARELLA: STRETCHING YOUR WAY TO SUCCESS

Mozzarella requires a stronger acid or rennet, and a slight manipulation of the curds, but the process is still incredibly fast. We will use citric acid for this quick version.

What You Need:

1 Gallon whole milk (non-UHT)

1.5 tsp Citric Acid (food grade, available online or at brewing stores) dissolved in 1/2 cup cool water

Rennet: 1/4 tablet or 1/4 tsp liquid rennet dissolved in 1/4 cup cool water (Vegetarian rennet is widely available.)

Thermometer

Large stainless steel pot

Slotted spoon

Gloves (for handling hot curds)

The Steps (The Stretch Test):

1. Prepare the Milk: Pour milk into the pot. Add the dissolved citric acid mixture and stir thoroughly.

2. Heat for Curd Formation: Begin heating the milk slowly. As you heat, you will notice small curds starting to form. Continue heating until the milk reaches 90°F (32°C).

3. Add Rennet: Remove the pot from the heat. Gently stir in the dissolved rennet solution using an up-and-down motion for 30 seconds.

4. Set the Curd: Cover the pot and let it sit for 5 to 10 minutes until the curd is firm enough to cut. Test this by inserting a knife and pulling the cut apart—it should separate cleanly.

5. Cut and Reheat: Cut the curds into 1-inch cubes right in the pot. Return the pot to low heat and slowly raise the temperature to 105°F (40°C), stirring gently every few minutes. This releases more whey.

6. Separate: Line a colander with cheesecloth and scoop the curds into it, allowing all the whey to drain away. Press lightly to encourage drainage.

7. The Hot Water Stretch: This is the key. Heat a separate pot of water (or the remaining whey) to 175°F (80°C). Dip the curds into the hot water for about 30 seconds, then lift them out with a slotted spoon. Using gloved hands, begin folding and stretching the curds. Repeat the dip and stretch until the cheese is smooth, shiny, and pliable like taffy.

8. Form and Chill: Shape the hot, pliable cheese into small balls. Submerge immediately in a bowl of ice water to set the shape. Your fresh, soft mozzarella is ready for slicing in minutes.

FERMENTATION MAGIC: THE 3 ESSENTIAL MICROBES THAT BOOST YOUR IMMUNITY (THE FAIL-PROOF, NO-MOLD METHOD FOR PERFECT SAUERKRAUT EVERY TIME)

If the garden is where you create life, the kitchen is where you elevate it. Preservation is critical, but fermentation is pure alchemy. It takes the humble vegetable and transforms it into a living, immune-boosting superfood that lasts for months without refrigeration (as long as it's kept cool).

This process is fueled by Lactic Acid Bacteria (LAB). These beneficial microbes, often naturally present on the surface of organic vegetables, are the true engineers of your pantry. They consume the natural sugars in the vegetables and produce lactic acid, which creates a highly acidic environment where bad bacteria (like mold and pathogens) cannot survive. The result is a vibrant, vitamin-rich food that completely overhauls your gut health—the foundation of all immunity.

The benefits are immediate: better digestion, increased absorption of nutrients, and a powerful supply of probiotics far superior to any capsule you can buy.

MINIMAL EQUIPMENT FOR MAXIMUM IMPACT

You do not need specialized pottery or expensive airlocks to start. All you truly require are three items:

1. Wide-mouth mason jars (quart or half-gallon size).
2. Non-iodized salt (critical—iodine interferes with fermentation).

3. A weight (a clean, boiled river rock, a small glass dish, or specialized ceramic weights work perfectly).

THE FAIL-PROOF, NO-MOLD SAUERKRAUT METHOD

This method eliminates the guesswork and the risk of mold, ensuring a perfect batch every time. We are using salt and pressure to draw the water out of the cabbage, creating its own protective brine.

Step 1: Prep and Slice. Select one medium to large head of green cabbage. Remove the outer leaves (save these—they will be your mold-stopper). Core the cabbage, then slice it thinly, ideally using a mandolin or food processor for uniform shreds.

Step 2: The Salting. Place the shredded cabbage in a large, non-reactive bowl (stainless steel or ceramic). For every medium head of cabbage (about 3 pounds), add 1.5 tablespoons of non-iodized sea salt.

Step 3: The Massage (The Game Changer). This is the most crucial step. Using clean hands, begin massaging and squeezing the salt into the cabbage shreds. Be aggressive. You are mechanically breaking down the cell walls to release the vegetable's natural water. This takes time—about five to ten minutes of vigorous squeezing. When you finish, you should see a significant pool of brine at the bottom of the bowl.

Step 4: Pack and Submerge. Take handfuls of the salty cabbage and pack them tightly, layer by layer, into your mason jar. Use a wooden spoon or a tamper to press down hard after each layer. You want to eliminate any trapped air pockets. The brine you created in the bowl must rise above the cabbage layer as you pack.

Step 5: The No-Mold Secret (The Weight). Once the jar is packed and the liquid brine is above the cabbage, it is essential to keep all solids submerged. If cabbage floats above the brine, it is exposed to oxygen and will mold. Take one of the large outer leaves you reserved earlier and fold it to create a cap. Place it over the shreds, pressing it below the brine level. Then, place your weight directly on top of the leaf. This keeps the shredded cabbage safely below the water line.

Step 6: The Wait. Screw the lid on loosely (if using a standard jar) or apply an airlock lid. The fermentation process produces carbon dioxide, and this gas must be allowed to escape. Place the jar in a cool, dark place (ideally 60–70 degrees F). Wait 7 to 14 days. You will see bubbles forming —that's the LAB working hard. Taste it after a week; when it reaches a pleasant level of sourness and crunch, seal the lid tightly and move it to the refrigerator.

This simple technique—the combination of salt, pressure, and the submersion weight—can be applied to almost any firm vegetable, including carrots, green beans, or spicy pepper mixes. You are not just saving food; you are creating food that actively works to keep your family healthy and resilient.

HERBAL MEDICINE CABINET: DITCH THE PHARMACY: GROWING YOUR OWN 12-ITEM NATURAL FIRST AID KIT (HOW TO INFUSE HEALING TINCTURES AND SOOTHING SALVES)

WHAT IF THE MOST POTENT FIRST AID KIT YOU OWNED WASN'T PURchased from a sterile box but grew right outside your back door? This is the ultimate act of self-sufficiency: knowing you can address minor injuries, soothe upset stomachs, and bolster immune systems without relying on an external, vulnerable supply chain.

We are not aiming to replace emergency medical care, but we can absolutely replace 90% of the minor remedies we currently buy. The pharmacy is now in your garden.

The foundational herbs for a self-sufficient medicine cabinet are chosen for their ease of growth and powerful, broad-spectrum healing properties. Start small with these highly effective plants:

Chamomile: The ultimate nervous system ally. Used as a tea, it calms the gut, aids sleep, and soothes high-strung energy. It grows quickly and is prolific.

Peppermint: A digestive powerhouse. A strong infusion (tea) can quickly settle indigestion, relieve tension headaches, and open sinuses.

Calendula: The skin savior. The bright orange petals are profoundly anti-inflammatory and antiseptic, making them perfect for infusing into oil for salves and balms for cuts, scrapes, and dryness.

Plantain (not the banana kind): This common backyard weed is a phenomenal drawing herb. Chewed or crushed and placed directly on a bee sting or bug bite, it quickly pulls out irritants and reduces swelling.

Lavender: Anti-microbial, calming, and fragrant. Used in salves for burns or simply added to baths for relaxation.

Echinacea: Essential for immune support. Both the roots and the above-ground parts (flowers and leaves) can be used to create immune-boosting tinctures during cold and flu season.

How to Infuse Healing Tinctures

Tinctures are concentrated liquid extracts—the essential compounds of the herb suspended in alcohol. The process is deceptively simple. This is known as the "folk method" and requires no specialized equipment beyond a mason jar.

1. Harvest and Chop: Select the desired fresh or dried herb (such as Echinacea root for immune boost). Chop the material finely to expose more surface area.

2. Pack the Jar: Fill a clean glass jar about halfway with dried herbs, or three-quarters full if using fresh herbs.

3. Add the Solvent: Cover the herbs completely with a high-proof alcohol—usually 80-proof (40%) vodka or higher. The alcohol draws out the medicinal compounds that water cannot.

4. Infuse: Seal the jar tightly. Place it in a cool, dark spot. Shake it vigorously once a day for 4 to 6 weeks.

5. Strain and Store: After the infusion period, strain the liquid through cheesecloth, squeezing out every drop. Label the tincture clearly with the herb, the date it was made, and the alcohol used. Store in small, dark dropper bottles.

Dosing Tinctures: Start low. A typical dose for an immune-support tincture is 30–60 drops (one to two dropper-fulls) taken three times a day at the onset of symptoms.

Creating Soothing Salves and Balms

Salves are topical treatments—the herb's medicinal properties infused into oil and then solidified with wax. This is the cornerstone of a natural first aid kit, excellent for moisturizing and healing skin issues.

1. The Oil Infusion: Fill a jar loosely with dried herbs (like Calendula petals or Plantain leaves). Cover completely with a carrier oil, such as olive oil or sunflower oil. To speed up the process, you can place the jar in a double boiler over very low heat for 2–3 hours. Alternatively, let it infuse in a sunny window for 2–3 weeks (the solar infusion method).

2. Strain the Oil: Strain the oil through cheesecloth until no plant material remains. This is your potent, medicated oil.

3. Thicken with Wax: Measure your strained oil. For every cup of oil, you will need approximately one ounce of grated beeswax (this can vary based on desired firmness).

4. Melt and Combine: Gently heat the oil and the beeswax together in a double boiler until the wax is completely melted and incorporated into the oil.

5. Pour and Cure: Pour the liquid salve mixture immediately into clean, sterilized tins or small jars. Let it cool completely until solid.

The resulting Calendula salve will be your go-to balm for treating dry, cracked hands and minor skin abrasions, offering antiseptic and anti-inflammatory relief straight from the garden.

Always label everything you make. Know your source, know your herbs, and treat your garden with the respect due to the world's oldest pharmacy.

CHAPTER EIGHT

THE $14,000 ANNUAL SECRET: TURNING YOUR GARDEN SURPLUS INTO TAX-FREE CASH AND RADICAL DEBT REDUCTION

YOU HAVE DONE THE HARD, FOUNDATIONAL WORK. YOU HAVE SEcured your water, built your soil, and learned to raise meat and grow vegetables efficiently. You have achieved food security, which is the ultimate form of insurance. But there is a massive financial upside to this journey that 99% of people miss, simply because they never sit down to calculate the numbers properly.

We started this book focused on saving money by eliminating dependence on the unstable supply chain. Now, we shift gears to focus on generating significant wealth and achieving radical debt reduction through the high, often hidden, value of your surplus.

The figure we aim for—the $14,000 annual secret—is not an arbitrary number. It is a conservative calculation based on the cumulative effect of three factors:

1. The real retail cost of all the food you no longer purchase (savings).

2. The premium price you can command for hyper-local, artisan surplus goods (income).

3. The strategic use of bartering and agricultural tax loopholes (leverage).

Your homestead, no matter how small, is no longer a hobby. It is a highly efficient, high-yield asset. It is an investment that provides an immediate, measurable return on time and labor, completely insulated from market volatility.

Think about the grocery store. When you buy a dozen eggs, you pay for the feed mill, the industrial farm's overhead, the transportation, the refrigeration, the centralized processing, the distributor, and the retailer. When you use your own homestead eggs, you eliminate every single one of those markups. That eliminated cost becomes pure, immediate equity.

Furthermore, the quality of your product—that deep orange yolk, that unblemished zucchini, that hand-cured bacon—commands a premium price in local markets that dwarf the cost of industrial alternatives. People are desperate for verifiable quality. You have that quality.

This chapter is the pivot point. It is where your resilience becomes profitable. We are going to put aside the fear of dependence and exchange it for the confidence of financial self-mastery. We will define the true, quantifiable value of your labor and product, learn to trade that value efficiently within your community, and leverage every opportunity to transform your surplus into cash and accelerated financial freedom. Let's look at the mathematics of independence.

CALCULATING YOUR SAVINGS: SHOCKING MATH: WHY A DOZEN HOMESTEAD EGGS IS WORTH $12.50 (ACCURATELY CALCULATING THE TRUE VALUE OF YOUR LABOR)

THE BIGGEST FINANCIAL QUESTION LOOMING OVER ANY SELF-SUFFICIENCY project is deceptively simple: Does this actually save me money?

If you only compare the grocery store price of a dozen conventional eggs ($3.00) to the immediate, tangible costs of your homestead eggs ($2.50 for feed, grit, and water), the initial math looks discouraging. This is where most people quit tracking their financial independence before they even start.

The crucial mistake is comparing apples to rotten oranges. You must employ True Cost Accounting. You are not producing a factory egg. You are producing a hyper-premium, resilient asset.

Let's calculate the real value of one dozen eggs produced on your homestead:

1. Base Cost (Feed and Infrastructure): $2.50

This covers the direct input costs: quality organic feed, water, and a fair annual depreciation charge on the coop structure and feeders.

2. Quality Premium (Nutritional Value Equivalent): +$3.50

Store-bought eggs that truly match the quality of your fresh, pastured, deep-orange-yolk eggs—organic, soy-free, free-range—easily cost $6.00 to $7.00 per dozen. Using a comparable retail price of $6.00, your savings on quality alone is $3.50. You are paying for food that actively improves your health.

3. Resilience and Security Value: +$2.00

This is the cost of your insurance policy. When food inflation spikes or the supply chain breaks, the guaranteed presence of fresh, reliable protein in your backyard is financially critical. Assigning a conservative $2.00 value for this peace of mind and certainty is reasonable.

4. The Value of Your Labor: +$4.50

You spent time. Time is money earned or time saved. Assume you spend approximately 15 minutes per week tending to five laying hens (gathering, feeding, cleaning). If we value your specialized, self-sufficient labor at a modest $18.00 per hour (a standard living wage), those 15 minutes of work equate to $4.50 in earned value. This is the money you are generating through production rather than spending through consumption.

5. Convenience and Transportation Savings: +$1.00

You eliminate the cost of gas, the wear and tear on your car, and the time spent driving to the grocery store.

Total Realized Value Per Dozen:

$2.50 (Base Cost) + $3.50 (Quality Premium) + $2.00 (Security) + $4.50 (Labor Value) + $1.00 (Convenience) = $13.50.

Suddenly, those eggs you gathered are worth $13.50 on the open market, or rather, they represent $13.50 in realized wealth to your household economy. If your conventional grocery bill for a dozen conventional eggs was $3.00, you have just pocketed an effective savings/income of $10.50. This is the shocking math that proves your labor is paying off.

THE CANNED TOMATO METRIC

This principle applies equally to preserved goods. A jar of high-quality, organic canned crushed tomatoes costs $4.50 in a store. If you grow, harvest, and can your own, the physical input cost (seed, water, electricity for the canner) might be only $0.80 per jar.

Applying the True Cost formula:

Retail Equivalent Value: $4.50

Labor Value (Canning/Processing): $3.00 (valuing 10 minutes of your processing time)

Security Value (Shelf-stable, supply-chain independent): $1.00

Total Realized Value: $8.50 per jar.

If you produce 100 jars of tomatoes in a season, you have effectively generated $850 in measurable wealth that never has to cycle through a wage-earning job or a retail checkout line.

This rigorous financial tracking is not optional. It is critical for two reasons: First, it quantifies and validates your hard work, transforming it from a hobby into a productive enterprise. Second, it shifts your perspective from seeing your garden as a potential expense to recognizing it as a substantial wealth generator and the most reliable insurance policy against instability.

Start meticulously tracking your inputs (seed cost, feed cost) and your outputs (dozens of eggs, pounds of canned goods). Once you see the measurable return on your labor, you gain the financial confidence needed to scale your operation and secure your family's future.

BARTERING AND TRADE: THE SUGARMAN BARTER PRINCIPLE: HOW TO GET $300 WORTH OF BUTCHER LABOR BY TRADING 10 JARS OF PICKLES (MASTERING NON-MONETARY WEALTH)

Money is a tool. But it is not the only currency that matters on a resilient homestead. When you are focused on self-sufficiency, your time, your specialized labor, and your high-quality products become far more valuable than the paper bills sitting in your wallet.

The goal of true resilience is to reduce reliance on centralized systems, and this includes reducing dependence on strictly monetary transactions. Bartering is the key to mastering non-monetary wealth, ensuring that even if the cash economy stutters, your household keeps running.

The difference between a successful barter and a failed one is perceived value. This is the core of what we call the Sugarman Barter Principle.

The traditional mindset dictates that if you want a professional butcher to process a pig—a service that might cost $300 cash—you must pay cash. The Sugarman Principle flips this. Instead of cash, you approach the butcher (or mechanic, or carpenter) with an item of such obvious quality and perceived scarcity that it becomes a more desirable form of payment than money.

Consider the butcher exchange. You offer 10 jars of expertly preserved, spicy dill pickles or maybe 5 pounds of your homestead's specialty artisan bacon. The cash cost of those ingredients might be $50, but the butcher recognizes that these items:

1. Cannot be easily purchased in a store (they are unique).

2. Represent a skill the butcher does not possess (preservation).

3. Save the butcher time and provide high-quality food for their family.

For a busy specialist, trading $300 worth of his time for a winter's supply of high-end preserved food is often a better deal than receiving cash. It simplifies his budget and secures a luxury item he wants. This is the Sugarman Barter Principle in action: maximizing the perceived value of your specialized surplus to acquire specialized labor you cannot easily do yourself.

What are your most valuable bartering assets?

1. Specialty Preserves: Jams, jellies, pickles, cured meats, or even unique fermented foods. These are high-value because they represent time and expertise.

2. Premium Protein: Excess eggs from grass-fed chickens, or specialty cuts of ethically raised rabbit or goat meat.

3. Labor and Skill: Are you proficient in welding? Do you know how to lay simple brickwork? Trade an afternoon of welding a neighbor's gate for a sack of their specialty potatoes.

4. Starts and Seeds: If you're a master seed saver, your unique heirloom seed stock is highly valued by other growers.

THE RULES OF EFFECTIVE BARTERING

Bartering is built on reputation. Unlike a cash transaction where you never see the seller again, bartering is fundamentally about establishing a long-term, trustworthy relationship with your local community.

Rule 1: Never Negotiate Down on Quality. Whatever you offer must be the absolute best representation of your craft. If you trade eggs, they must be fresh that day. If you trade soap, it must be perfectly cured. This consistency builds immense trust and ensures repeat trades.

Rule 2: Clearly Define the Scope. Before any exchange, define exactly what is being traded for what. Is the 10 jars of pickles for a full pig breakdown, or just half? Define the labor hours versus the goods. Ambiguity kills trust faster than anything.

Rule 3: Start Local. Your best trading partners are other local homesteaders who also have a specialized surplus. A dairy farmer has excess milk (which you need for cheese); you have excess beef or garden produce (which they need for variety). These exchanges form the basis of a resilient local food web, ensuring that everyone's basic needs are met without relying on a broken supply chain.

By mastering bartering, you acquire goods and labor without depleting your hard-earned cash, turning your garden surplus into a powerful, tax-free asset. You are trading what you have in abundance for what you need most—a fundamental step in true self-sufficiency.

—★-★-★—

SELLING THE SURPLUS: THE 'ROADSIDE STAND CASH COW': PRICING STRATEGIES THAT MAKE PEOPLE WAIT IN LINE FOR YOUR $6 ZUCCHINI BREAD

THE DIFFERENCE BETWEEN SELLING $1 ZUCCHINI AND SELLING $6 zucchini bread is the Transformation Tax—the value you add through skill, time, and presentation. Your goal in selling surplus is not to move volume at cost, but to monetize the exceptional quality and trust you have built into your product.

PRICING FOR PREMIUM VALUE

If you price your product based on the supermarket's price, you will fail. The supermarket is built on volume and cheap labor; you are selling scarcity and quality. Your true competitors are high-end organic markets or specialized farmer's markets, not the mass retailer down the street.

1. The 2x Organic Rule: A good starting point is to price your goods at 1.5 times to 2 times the price of the equivalent organic item at the local grocery store. This immediately communicates that your product is superior. If organic eggs are $5/dozen, your homestead eggs should be $7.50 to $10.00. Customers who seek out roadside stands are looking for quality and local connection, and they expect to pay for it.

2. The Labor Multiplier: When selling processed goods (like that $6 zucchini bread or $8 jars of jam), you must factor in your labor time at a rate you respect ($25/hour is fair). Do not simply tally the ingredient cost. If it takes you one hour to make four loaves of bread, you must recover $6.25 in labor per loaf, plus ingredients, plus profit. This approach ensures you are financially rewarding your highly skilled effort.

3. The Anchor Price Strategy: Strategic placement of high-value items makes your standard items look like a great deal. If you want to sell a lot of your $6 zucchini bread, also offer a small batch of "Spiced Winter Squash and Maple Loaf" for $12. The $6 bread suddenly seems accessible and reasonable by comparison, boosting overall sales.

PRESENTATION: SELLING THE EXPERIENCE, NOT THE PRODUCT

Your display must communicate the high value that justifies the premium price. Do not stack your produce in old cardboard boxes.

1. Clean, Consistent Aesthetics: Use clean baskets, wooden crates, or small, matching containers. If you use plastic bags for items like salad greens, ensure they are high-quality, clear, and uniformly tied with twine or a neat label. Presentation is silent marketing.

2. The Power of the Label: Customers are buying your story. Even on a simple honor-system roadside stand, every item should have a hand-written or stamped label that details the unique selling proposition: "Heirloom Cherokee Purple Tomatoes – Picked 6 AM Today," or "Pastured Duck Eggs – Richer Yolks for Baking." This establishes the quality and origin immediately.

3. Sell the Processed Goods: The "Cash Cow" items are the ones where you have added value. Everyone has surplus zucchini, but few have high-quality, vacuum-sealed Zucchini Bread or frozen pesto made from homestead basil. Processing converts low-value surplus into high-margin artisan goods. Prioritize selling jams, baked goods (where permissible by local cottage food laws), infused honeys, and specialty rubs.

SETTING UP THE ROADSIDE STAND

THE KEY TO A SUCCESSFUL ROADSIDE STAND IS VISIBILITY, SAFETY, and trust.

1. Location and Visibility: Your stand must be highly visible from both directions, with clear signage posted early. Crucially, ensure there is a safe, stable shoulder or driveway where cars can pull completely off the road without blocking traffic. If customers feel unsafe stopping, they will drive past.

2. The Honor System: For small, unmanned stands, the trust-based honor system works surprisingly well, especially in rural or semi-rural areas. Provide a secured cash box (a lockbox bolted down is ideal) and clear instructions for payment. To facilitate sales in the digital age, include a large,

clear sign with your preferred digital payment method (Venmo, PayPal, CashApp QR codes). Trust often breeds reciprocal honesty and encourages sales.

3. Consistency and Variety: Customers shop where they know they will find what they need. If you can only staff the stand on Saturdays, make it known that Saturday is the day. Furthermore, offer variety. A handful of tomatoes is okay, but a stand offering tomatoes, fresh herbs, bread, honey, and soap is a destination. If your eggs run out, don't just leave the stand empty; fill it with flowers or extra preserved goods.

—★-★-★—

TAX BREAKS AND LICENSING: THE FARMER'S TAX LOOPHOLE: USING AGRICULTURAL DEDUCTIONS TO KEEP MORE OF YOUR HARD-EARNED MONEY (EVEN ON A QUARTER ACRE)

THE FINANCIAL BENEFITS OF INTENTIONAL FOOD PRODUCTION EXtend far beyond the savings at the checkout counter. Once you begin treating your homestead activities—even the quarter-acre garden—with the formality of a business, you unlock powerful mechanisms designed to reward American agriculture: tax deductions.

It is crucial to understand the distinction the IRS makes between a "hobby" and a "for-profit business." To claim substantial deductions using Schedule F (Profit or Loss From Farming), your operation must demonstrate a reasonable expectation of profit in three out of five consecutive years. Even if you start small, meticulous record-keeping (Chapter 1.3) from day one is the evidence you need to prove your operation is indeed a business venture, not just a weekend pastime.

What can be deducted? Nearly anything necessary for production. This can include the cost of seeds, fertilizers, livestock feed, fencing materials, the dedicated portion of electricity used for your coop or greenhouse, repair costs for tools, and even mileage to the feed store or farmers market. If you install a rainwater catchment system or a solar array primarily intended to irrigate crops or power livestock waterers, a portion of that expense may also be deductible. These agricultural deductions can dramatically reduce your taxable income, effectively allowing the government to subsidize your path to self-sufficiency. Always consult a Certified Public Accountant (CPA) who specializes in farm or small business taxes to ensure compliance.

LICENSING AND SELLING YOUR SURPLUS

Before you open your roadside stand or commit to selling your artisan jam, you must investigate local licensing requirements. Regulations governing food sales are often layered, involving federal, state, and county health departments.

The most important legal gateway for the small producer is the existence of Cottage Food Laws. These state-level laws are specifically designed to help small producers sell low-risk foods (like baked goods, certain candies, jams, jellies, and dried herbs) made in a home kitchen, bypassing the massive expense of a commercial kitchen lease. However, these laws usually come with strict stipulations:

1. Annual Revenue Caps: Most states cap the total revenue you can earn from cottage foods (often between $25,000 and $50,000).

2. Labeling Requirements: You must often label the product clearly stating it was "Made in a home kitchen not subject to state inspection."

3. Sales Location Restrictions: Sales are usually restricted to direct-to-consumer venues like farmers' markets, roadside stands, or online—not wholesale to grocery stores.

For raw produce and eggs, the requirements are often less strict, but still require diligence. Selling eggs, for example, may require specific washing protocols, refrigeration standards, and carton labeling (including the producer's name and address) if you plan on selling more than a handful of cartons weekly.

The takeaway is simple: Do the research first. A simple phone call to your county health department and planning department will clarify what products you can legally sell and the necessary permits required for your small-scale operation to transform into a legitimate, tax-advantaged source of income.

CHAPTER NINE

RESILIENCE ENGINEERING: HOW TO MAKE FAILURE YOUR GREATEST ASSET (THE LONG GAME OF FREEDOM AND SELF-MASTERY)

THIS BOOK BEGAN WITH A CHALLENGE: TO FIND FREEDOM FROM fragility. We have covered the blueprints for soil, water, structure, and savings. We have quantified the value of every egg and every preserved jar of tomatoes. You now hold the practical map to self-sufficiency.

But the blueprints are not the whole story.

The truth about homesteading—whether you are working half an acre or ten—is that the technical skills are only half of the equation. The other half is purely mental. It is about emotional endurance. It is about understanding that the biggest disaster is not the failure itself, but the decision to quit when failure arrives.

Resilience is not a buzzword you print on a t-shirt. It is the ability to walk outside the morning after the hail storm and start clearing the damage, knowing that the earth does not care about your schedule or your effort. It operates on its own terms.

When the grid goes out, we rely on the solar array we built. When the store shelves empty, we rely on the root cellar we dug. But when the fence fails, when the blight hits the entire potato crop, when the new hive swarms and flies away—what do we rely on then?

We rely on the quiet, often bruised, conviction that we are builders, not quitters.

We learn, very quickly, that this life is not a linear climb toward perfection. It is a slow, messy, beautiful spiral of constant improvement. Every major setback—the sudden loss of livestock, the inexplicable failure of a harvest, the expensive breakdown of a crucial pump—is a tuition payment in the only school that truly matters: the school of knowing exactly what you can handle.

This final chapter is the emotional engineering required to sustain this life for the long haul. Because securing your family's food supply is a lifelong commitment, a quiet revolution fought one season at a time. It demands honesty about our limitations, wisdom about our pace, and connection to the people who share the journey.

We are entering the long game now. The game of freedom and self-mastery. And in this game, failure is not the opposite of success; it is the most valuable asset you possess. It is the curriculum for tomorrow's victory.

——★–★–★——

DEALING WITH FAILURE: THE $500 MISTAKE I MADE: HONEST ANECDOTES ABOUT TOTAL GARDEN COLLAPSE (AND THE 3-STEP PROTOCOL TO BOUNCE BACK IN 24 HOURS)

The dirt itself is the toughest teacher you will ever meet. It doesn't care about your goals or your schedule. It simply enforces physics and biology.

If you are committed to this life, you must accept failure as a required step. You are going to watch things die. You are going to waste time, money, and emotional energy. The difference between a homesteader and someone who gave up after the first season is how they handle the collapse.

I made a terrible mistake in my second year—a costly one. I wanted chickens, but I was fixated on doing it cheaply and fast. I built what I thought was a fantastic mobile chicken run using thin wire fencing, the kind you might use for a temporary garden bed. I reasoned that since I lived in a semi-urban area, predators wouldn't be a huge concern. I was wrong. The local fox had a very different idea about free lunch.

One early morning, I walked outside to silence. The fox had easily chewed through the thin wire—not hardware cloth, which is essential—and wiped out my entire flock of 15 beautiful laying hens. I had lost the birds, the feed investment, the future eggs, and the immense pride I felt in my self-sufficiency. In one night, it was a loss of over $500, not including the value of my time.

I sat on the ground and felt a profound, heavy anger. This was the moment I could have quit. I could have sold the structure, cursed the fox, and driven back to the grocery store. Instead, I grabbed my notebook. I diagnosed the failure immediately: Inadequate security due to material substitution. Saving $50 on proper hardware cloth cost me $500 in capital loss. The lesson was expensive, but the lesson stuck: Never compromise on security or foundations.

These setbacks are your tuition fee. Maybe you forgot to sterilize your jars, and 10 pounds of preserved peaches turned into fermented goo. Maybe you misjudged the spring weather, and a late frost killed all 72 of your meticulously started tomato seedlings. Perhaps the rain was too heavy, and your prize-winning zucchini crop floated away. It happens.

The key is not to avoid failure, but to implement the 3-Step Protocol to bounce back in 24 hours. This protocol is designed to short-circuit the paralysis and the desire to quit.

STEP 1: STOP THE SPIRAL. (THE 10-MINUTE RULE)

ALLOW YOURSELF TEN MINUTES TO RAGE, COMPLAIN, OR FEEL FRUSTRATED. Literally set a timer. When the timer goes off, the emotion stops. You must switch immediately from emotional reaction to analytical assessment. What exactly failed? Was it the system, the execution, or the environment? Stop trying to fix it right this second. Just stop the damage. If a fence fell, block the opening with a trash can. If jars spoiled, throw them out. Stop the bleeding.

STEP 2: WRITE THE LESSON

PULL OUT YOUR NOTEBOOK (THE SAME ONE YOU USE FOR TRACKING your yield). Do not minimize the failure. Document the specific conditions, materials, and steps that led to the collapse. This is not about assigning blame; it is about gathering data. For my chicken failure, the lesson was: "Wire fencing is not predator proof; must use 1/2-inch hardware cloth buried six inches deep." This lesson is now enshrined in your manual. You cannot make the same mistake twice if it is documented and accepted as a fact.

STEP 3: IMPLEMENT THE TINY FIX

THE MOMENTUM OF A MAJOR FAILURE CAN CRUSH YOUR SPIRIT. To restore confidence, you need an immediate, small win. Don't try to rebuild the entire collapsed garden bed immediately. Instead, do the smallest, most essential task: Clean the tools, order the replacement hardware cloth, or simply transplant three healthy seedlings that survived. Choose a task that takes less than an hour and guarantees a minor success. This tiny fix proves you are still in control. It restores momentum.

The self-sufficient life is defined by resilience, not perfection. Every time you pivot successfully after a collapse, you are not just fixing a garden; you are engineering a stronger self. You are gaining the mastery that no book or YouTube video could ever teach you. You paid the tuition. Now use the knowledge.

—★-★-★—

EXPANDING SLOWLY: THE 'ONE-THING-AT-A-TIME' RULE: HOW TO AVOID BURNOUT AND MASTER BEES, FRUIT TREES, OR GOATS WITHOUT LOSING YOUR MIND

THE MOMENT YOU TASTE THAT FIRST TRULY SUCCESSFUL HARVEST—the heavy tomatoes, the perfectly laid eggs—you feel an irresistible urge. You think: If I can do this, imagine what I could do with two goats! Or a couple of beehives!

That moment of victory is also the single most dangerous moment for the new homesteader. It is where burnout begins.

You are now working toward resilience, not just production. Resilience comes from systems that work almost on autopilot. When you try to stack a new, complex system—like managing the delicate ecosystem of a beehive—on top of an existing system that hasn't yet stabilized (like keeping your chickens alive through winter), both systems suffer. You dilute your focus, stress your resources, and guarantee a failure that is often traumatic enough to make you quit entirely.

This is why we must adopt the "One-Thing-At-a-Time" Rule (OTAT).

The OTAT Rule states that you must achieve repeatable, predictable mastery in three core areas before introducing a single new element that requires significant daily management or specialized skill.

What does mastery look like? It means:

1. Consistency. You have survived at least one full calendar year (ideally two) using your current systems, from planting to preservation, with predictable yields.

2. Protocol. You have written down the standard operating procedure for every essential task (composting recipe, chicken health check, preservation method). You are executing the task, not guessing at it.

3. Resilience. You have handled one major system failure (e.g., a garden pest invasion, a chicken illness, a water system freeze) and successfully recovered using your learned protocols.

If your garden is still a crisis zone every August, you are not ready for a micro-dairy goat that needs milking twice a day and veterinary oversight. If you are throwing away last year's spoiled canned goods, you are certainly not ready to manage fruit trees, which flood your schedule with an enormous, perishable yield all at once.

Consider the logical expansion sequence. You move from the least complex and highest-yield systems to the most complex and long-term ones:

Phase 1: Soil and Garden (Compost, Raised Beds, Seed Starting).

Phase 2: Preservation and Water (Canning, Freezing, Rain Catchment).

Phase 3: Poultry (Chickens, Ducks—high return, small daily commitment).

Phase 4: Perennials (Fruit Trees, Berry Bushes—long-term investment, slow return).

Phase 5: Specialist Systems (Bees, Goats, Rabbits—require intense, focused study).

Before adding a new element like bees, ask yourself: Will I have the concentrated time needed in April and May for swarm management and hive checks? Or will I be distracted by my failing fence line and a spring garden that is two weeks late?

To avoid burnout, you must treat your self-sufficiency journey like building an architectural foundation. If the first layer of concrete (your soil and garden) is wet and shaky, adding a second story (goats) is sheer recklessness. Wait until the concrete is dry. Wait until the system runs itself. The long game demands patience. The reward is not just more food, but a deeper, calmer competence that anxiety cannot touch.

TEACHING THE NEXT GENERATION: THE LEGACY PROTOCOL: INSTILLING RESILIENCE AND RESPONSIBILITY IN YOUR KIDS BY GIVING THEM THEIR OWN PLOT (THE GREATEST GIFT YOU CAN GIVE)

WHO WILL HOLD THE KNOWLEDGE WHEN WE ARE GONE? THIS IS THE most profound question we face as homesteaders. We invest countless hours in learning how to secure food, water, and shelter, yet these skills are useless if they die with us. The greatest dividend of self-sufficiency is not the food we harvest, but the competence we instill in the next generation.

We are not just teaching them to plant seeds; we are teaching them resilience. We are teaching them that failure is information, and that their capacity to adapt determines their survival.

The most effective tool in the Legacy Protocol is simple: give them agency. Give them dirt that is purely theirs.

THE OWNERSHIP PLOT

This doesn't require a dedicated acre. It can be a single 4x4 raised bed, a specific row in the garden, or even a couple of large containers on the patio. This is their domain. They choose the seeds—even if they want to grow ten varieties of novelty squash that you know won't work. The point is not maximum yield; the point is maximum engagement.

When the bed is theirs, the work is no longer a chore assigned by a parent, but a responsibility accepted by a steward.

Start the process young. A three-year-old can be the dedicated "Egg Fetcher," responsible for carefully gathering eggs and bringing them inside. A six-year-old can be the "Compost Manager," responsible for tossing the kitchen scraps and adding water. These tasks are small, but they demonstrate that the entire household system requires input from everyone.

As they grow, introduce true resource management. When your child is eight or ten, allow them to manage a small animal project. Perhaps a rabbit colony, or a few quail. This teaches them feeding cycles, cleaning schedules, and the solemn reality of life and death on the farm. They learn that the well-being of a creature depends entirely on their consistent effort.

THE ULTIMATE LESSON: PIVOTING FROM FAILURE

The real magic happens when things go wrong. And they will go wrong. Pests will decimate their chosen patch of cucumbers. They will forget to water the beans. They will harvest the carrots too late.

When this happens, do not rescue them. Instead, sit with them and ask: Why did this fail?

"The squash bugs ate everything."

"Okay, what do we know about squash bugs now? How do they reproduce? What eats them? What can we plant next year that they won't like?"

By analyzing the failure together, you transform disappointment into practical data. They learn that competence is not the absence of mistakes, but the ability to diagnose problems and pivot quickly. That feeling of recovering a garden after a setback is the root of true resilience. They realize they can handle a crisis because they have solved problems directly related to feeding themselves.

This connection to the earth instills a deep value system that consumerism cannot touch. When a child has spent four months tending a tomato plant, they do not casually throw away a half-eaten sandwich. They understand the cost—not in dollars, but in labor, sun, and time. This is the Legacy Protocol in action: securing their future by ensuring they possess the core skills, connection, and character needed to stand strong when the world around them becomes fragile.

FINDING YOUR TRIBE: WARNING: NEVER HOMESTEAD ALONE! THE HIDDEN NETWORK OF LOCAL FARMERS OFFERING FREE LABOR, EQUIPMENT, AND SPECIALIZED ADVICE

THE MYTH OF THE LONE, SELF-SUFFICIENT HOMESTEADER IS ONE OF the most romantic and most dangerous falsehoods in this lifestyle. While self-reliance is the goal, true resilience is found not in isolation, but in robust community. You cannot do this alone.

There will come a day when you need to lift something impossibly heavy—a beam for a barn, a full grain silo, or a newly processed half-carcass of meat. There will be days when you need specialized knowledge that no book can give you—why the blight hits the west side of your tomatoes but not the east, or the optimal timing for flooding the rice paddy in this particular microclimate. This institutional, hyper-local knowledge is priceless, and it is held exclusively by your neighbors, particularly those who have been working the same land for decades.

The hidden network of local farmers and seasoned homesteaders is your greatest asset, far more valuable than the most expensive piece of equipment. They hold the free labor, the specialized tools you could never afford, and the technical advice that will save you months of trial and error.

Finding and Activating Your Tribe

Your first step toward community engagement should be offering, not asking. You are establishing yourself as a reliable partner in the ecosystem. Start by visiting your local farmers' market frequently. Engage in conversation. Ask vendors questions about their techniques and their challenges. Never dismiss the value of a casual conversation at the feed store or the local co-op. These are the watering holes of the local tribe.

The local agricultural extension office is a necessary resource, often hosting meetings or classes where you can meet established growers. Similarly, seeking out regional farm-to-table events or agricultural fairs provides natural networking opportunities. Do not underestimate the power of social media groups dedicated to homesteading in your specific county or region. These groups often facilitate quick exchanges of advice and resources.

THE CURRENCY OF RECIPROCITY: LABOR AND EQUIPMENT SHARING

THE MOST IMMEDIATE BENEFIT OF YOUR NETWORK IS ACCESS TO EXpensive, specialized equipment. Consider a simple task like fence-post drilling. A heavy-duty gasoline-powered auger costs thousands of dollars and is rarely used more than a few days a year. Buying it makes no economic sense. But by connecting with local growers, you can often arrange to borrow it for a few days in exchange for a half-day of your labor, ten dozen eggs, or a winter's supply of preserved produce.

This is the principle of bartering labor, or the community work party, and it is critical for large projects. Fencing a large pasture, raising the frame of a shed, or processing multiple animals efficiently requires focused, fast labor. You cannot pay a professional crew for this work, but you can trade labor with five other homesteaders. You dedicate a Saturday to helping your neighbor plant a fall cover crop, and the following weekend, those five people show up at your place to help you pour concrete for your chicken coop expansion.

This system strengthens everyone. It eliminates the financial burden of specialized purchases, increases efficiency on critical jobs, and, most importantly, provides an essential safety net. When a disaster strikes—a barn fire, a sudden illness, or a broken pump—it is this local tribe that will rally to keep your operation running. Do not wait for a crisis to build these relationships. Start connecting today. Your independence is secured by your preparedness, but your resilience is secured by your community.

CHAPTER TEN

YOUR ANXIETY IS GONE. YOUR INDEPENDENCE IS SECURED. THE DAY YOU STOP BUYING YOUR OWN FOOD

You hold the blueprints now. The maps are laid out. The tools are defined. The budget is clear. You have absorbed the step-by-step instructions for building security, from amending your soil to safely preserving your harvest and generating income.

The fear of the unknown—the anxiety rooted in dependence—is replaced now by specific, actionable knowledge. But the knowledge in your mind does nothing for your plate.

The hardest part of this entire journey is truly starting.

The real work begins the moment you close this document and pick up a shovel. Go get your hands dirty. Go plant that first seed. Go build that first fence. Stop planning and start executing.

Remember, this transformation is not defined by the size of your land. It is defined by the depth of your resolve. It doesn't matter if you have one acre or just a small porch railing. Every small step toward self-sufficiency is a profound and measurable victory.

This is your life. This is your food. You are no longer relying on fragile, invisible systems. You have secured your independence. You have traded that nagging worry for the quiet, profound fulfillment that comes only from working the earth, feeding your family, and knowing precisely where your nourishment originates. That day you stop buying your own food, when the storage shelves are full and the garden is producing, is the day you achieve true freedom.

I am immensely proud of the commitment you have made to this path. Keep going. Keep learning from the soil, from the rain, and from the inevitable failures. And enjoy every quiet, fulfilled moment of your new, resilient life.

Go build your reality.

www.ingramcontent.com/pod-product-compliance
Lightning Source LLC
LaVergne TN
LVHW051044210926
848486LV00001B/13